Human Pedi

Proceedings of a Conference organised by the Galton Institute, London, 1998

Conference organisers John Timson,
Milo Keynes and John Peel

Edited by

Robert A Peel

PUBLISHED BY THE GALTON INSTITUTE

© The Galton Institute 1999

All rights reserved. No part of this publication may be reproduced or transmitted in any form or by any means without written permission or in accordance with the provisions of the Copyright, Designs and Patents Act 1988.

British Library Cataloguing in Publication Data

```
Human Pedigree Studies
1. Eugenics - Congresses 2. Human Genetics - Congresses
3. Genealogy - Congresses
I.  Peel, Robert A.
363.9'2
ISBN 0950406643
```

First published 1999 by The Galton Institute, 19 Northfields Prospect, Northfields, London SW18 1PE

Printed and bound in Great Britain by The Chameleon Press, 5-25 Burr Road, Wandsworth, London, SW18 4SG

Contents

Notes on the Contributors — v

Editor's Introduction
Robert Peel — vii

Genealogy: The Construction of Family Histories
Anthony Camp — 1

The Galton Lecture 1998: Eugenics: The Pedigree Years
Pauline Mazumdar — 18

Human Pedigrees and Human Genetics
Elizabeth Thompson — 45

A Brief History of the Pedigree in Human Genetics
Robert Resta — 62

Computers for Research, Storage and Presentation of Family Histories
David Hawgood — 85

Social, Ethical and Technical Implications of Pedigree Construction: What The Maps Tell Us About the Mapmakers
Robert Resta — 107

Index — 115

Notes on Contributors

Editor's Introduction

Robert Kark

Genealogy: The Construction of Family Histories

Anthony Smith

Australian Lecture 1996: Experiences — The Pedigree Years

Colin Thursday

Human defences and Human Genetics

Elizabeth Thompson

A Brief History of the Pedigree in Human Genetics

Robert Kark

Computers for Research, Storage and Presentation of Family Histories

David Hawksford

Social, Ethical and Technical Implications of Pedigree Construction: What the Maps Tell Us About the Mapmakers

Robert Kark

Index

Notes on the Contributors

Anthony J Camp, Former Director, The Society of Genealogists, London, and President of the Federation of Family History Societies

David Hawgood, Author and Publisher, London

Professor Pauline M H Mazumdar, Professor of the History of Medicine, Institute of History and Philosophy of Science and Technology, University of Toronto

Robert A Peel, President, The Galton Institute, London

Dr Robert Resta, Director, Genetic Counselling Services, Swedish Medical Center, Seattle, Washington

Professor Elizabeth Thompson, Professor of Statistics, University of Washington, Seattle, Washington

Editor's Introduction

Robert Peel

"There are many methods of drawing pedigrees and describing kinship, but for my own purposes I still prefer those that I designed myself."

- Francis Galton[1]

This book is based on papers presented at the Galton Institute's thirty-fifth annual conference held on 17 September 1998 at the Wellcome Institute for the History of Medicine. The conference was designed to examine the concept of the human pedigree in both its traditional and scientific forms and to assess the contribution of the eugenics movement to the development of this essential theoretical tool.

Eugenics was the intellectual link between Darwinism and human genetics in the post-1859 search for a theory of biological inheritance. Its major achievement was the perfection of the pedigree model as an analytic device. Developed in response to the problems posed by nineteenth century theories of heredity it was to become eugenics' unique gift to twentieth century human genetics; a construct as fundamental to that discipline as was the Periodic Table to chemistry.

The process of evolution through natural selection as proposed in the *Origin of Species* required, and promoted the search for, a concomitant theory of biological inheritance. Darwin's own attempts to formulate such a theory were muddled and unconvincing but, like all nineteenth century scientists, he was handicapped by being unaware of Mendel's classic paper - an original copy of which, its pages uncut, was found in his study at Downe following his death.[2]

Francis Galton played a major role in the quest for a theory of heredity in the decades preceding the rediscovery of Mendel's paper. He was already well informed on the subject through his independent researches on the inheritance of human ability which he reported, first in journal articles in 1865 and four years later in his book *Hereditary Genius*. Nevertheless, he acknowledged the influence of Darwin with whom he resolved to collaborate in his subsequent endeavours.

In his autobiography[3] Galton recalled: "I was encouraged by the new views [appearing in the *Origin of Species*] to pursue many inquiries which had long interested me and which clustered round the central topics of Heredity." W F Bynum describes Galton as "one whose intellectual life utterly hinged on Darwin's work" and suggests that "without the sense of process which Darwin's work provided the whole thrust of his research after the mid-1860s is unimaginable"[4]. Bynum also notes that no biographer of Darwin fails to mention the experimental critique that Galton offered of Darwin's theory of pangenesis.

Galton's collaboration with Darwin was conducted both privately through a steady exchange of letters (all of which survive) and publicly in the columns of *Nature*. It was a curious dialogue marked at the outset by sullenness on Darwin's part at Galton's criticism of pangenesis but overcome by the latter's tact, good humour and obvious unwillingness to offend his old friend and cousin. The result was a sustained and productive partnership which, by the time Mendel's paper was "rediscovered" in 1900, had considerably advanced inheritance theory, had anticipated a number of Mendel's ideas and provided others by which Mendelism could more easily be understood.

The observable facts that a theory of heredity was required to accommodate were simple and self-evident: *like produces like but with exceptions.* It was the exceptions that presented the real intellectual challenge and which, for different if converging reasons, were of especial significance to Darwin and Galton. For the former these exceptions provided the basis for evolutionary change; for the latter they furnished the evidence for the dominance of heredity over environment. "The sudden appearance of a man of great abilities in undistinguished families" and the occurrence of "mediocrities in [otherwise] illustrious families" were, in Galton's opinion, conclusive proof of the irrelevance of environmental influences.

Darwin's theory of pangenesis assumed that the carriers of hereditary information, which he termed "gemmules", originated in the cells of every part of the body, migrated to the reproductive organs and thus determined the character of subsequent offspring. Variation was the result of environmental influences affecting the parental cells the results of which would be transmitted via the gemmules. (Darwin was a Lamarckian to the end of his life). Galton determined to put Darwin's theory to the test. With the help of the staff of the London Zoo he transfused self-coloured grey rabbits with the blood of parti-coloured lop-eareds (effecting a fifty per cent exchange) immediately before mating the greys. The expectation, according to the theory of pangenesis, was that the resulting young would be brindled. In a two-year period Galton obtained more than eighty offspring in this way. All were uniformly grey.

But Galton did not as a consequence repudiate pangenesis. Whatever his personal doubts he was quickly made to realise that to do so would jeopardise his partnership with Darwin. Although long aware of Galton's experiments, when the first results were published in 1871 Darwin wrote a petulant letter to *Nature* denying, somewhat disingenuously, that the transmission of gemmules was dependant upon the circulation of the blood. Galton was not naturally averse to a good quarrel, as he showed in his involvement in the Stanley affair, but on this occasion he hastened to pacify Darwin. In a letter published in the next issue of the journal he conceded that he might have misunderstood Darwin's theory and ended: "Viva Pangenesis". It was a price he was prepared to pay in order to sustain the partnership.

The historians and philosophers of science may yet provide us with a detailed evaluation of this unique intellectual collaboration between two of the finest scientific minds of the nineteenth century. Those historians have, in recent years, challenged the popular legend of Mendel's "lost" paper. It was not, they suggest, unknown in the last decades of the nineteenth century; it was merely regarded as irrelevant - a contribution to hybridisation theory rather than to evolutionary biology.[5] There is no evidence that either Galton or Darwin was aware of Mendel's work, yet their collaboration produced some curious parallels with his ideas and methods.

It was at Darwin's suggestion that Galton undertook his experiments with sweet peas, providing weighed and measured seeds to friends who grew them on in different locations and returned the crop for analysis and comparison. From the results Galton was led to the complex mathematics of regression theory (he first called it reversion) which became a key concept in the biometric approach to genetics. In a further response to a query by Darwin, Galton "almost stumbled on Mendel's laws *a priori*".[6] Speculating on the problem of hybrids, he supposed that Darwin's gemmules paired off to form molecules and, if one was white and the other black, then in a large number of cases one quarter would always be white, one quarter black and one half grey. As John Maynard Smith observes, "What a tragedy that he did not apply his idea, not to the cells or "molecules" of the F1, but to the individual organisms in the F2!"[7] And with reference to Galton's subsequent investigations into the inheritance of eye colour, J H Edwards comments: "If Galton had split his grandparents by type he could hardly have failed to discover what was later termed Mendelism, and even if Mendel had discovered it twenty years earlier, an independent discovery expressed forcibly in clear English would have lead to Galtonism becoming the established term. We would then have had both Daltonism and Galtonism as descriptions of the atomic nature of chemical and biological processes."[8]

In collaboration with Darwin Galton thus anticipated Mendel's ratios as well as the concept of diploid inheritance. He also made the distinction between continuous variation ("properly so-called") and mutation ("sports") and between dominant and recessive traits (he called them "patent" and "latent"). According to the long established scholarly convention of priority these must now be counted amongst the many near misses with which the history of scientific discovery is littered. Galton's name is nevertheless secure amongst the founders of human genetic theory as the creator of its fundamental and indispensable theoretical tool - the pedigree diagram.

First presented in 1869 in *Hereditary Genius* (Figure 1), this diagram was adapted, in a depersonalised form, from a device that for five hundred years had only social and legal significance. As an analytical tool it accommodated all known empirical facts; it assumed no specific biological mechanism and indeed was immediately acceptable to both the Mendelians and the biometricians; above all it was unequivocally Galton's intellectual property - Mendel had not anticipated him here. The subsequent development of the pedigree as "the most commonly used tool in medical genetics"[9] is the subject of this volume.

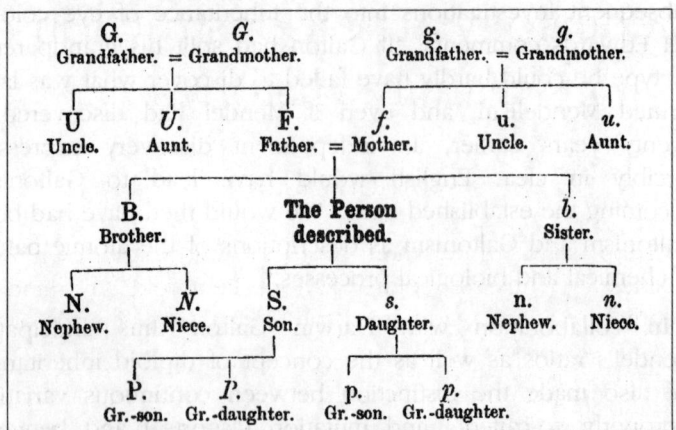

Figure 1: Galton's First Pedigree

The Galton Institute is indebted to the organisers of the conference on which this book is based and especially to the distinguished specialists who have contributed to it.

References:

[1] Galton, F (1889) *Natural Inheritance*, London: Macmillan

[2] Degler, Carl N (1991) *In Search of Human Nature*. OUP

[3] Galton, F (1908) *Memories of My Life*. London: Methuen

[4] Bynum, W F (1993) "The Historical Galton" in *Sir Francis Galton, FRS - The Legacy of his Ideas*, W M Keynes (Ed) London: Macmillan for the Galton Institute

[5] Olby, R C and Gautry P (1968) "Eleven References to Mendel Before 1900", *Annals of Science*, Vol. 24, quoted in Bowler, P J *The Mendelian Revolution*, London: Athlone Press, 1989

[6] Maynard Smith, J (1993) "Galton and Evolutionary Theory" in Keynes, *op. cit.*

[7] *ibid.*

[8] Edwards, J H (1993) "Francis Galton - Numeracy and Innumeracy in Genetics" in Keynes, *op. cit.*

[9] Resta, Robert G (1993) "The Crane's Foot: The Rise of the Pedigree in Human Genetics", *Journal of Genetic Counselling*, Vol. 2, No. 4.

Genealogy: The Construction of Family Histories

Anthony Camp

"Thou silly fellow, thou dost not know thy own silly business". Thus the soldier Earl of Pembroke, a friend of George the Second, addressed the Garter King of Arms of his day.

It is not surprising that a later Garter King of Arms and perhaps the greatest genealogist of this century, Sir Anthony Wagner, should have begun a talk to the Society of Genealogists in 1961 with the defensive words, "No true genealogist needs to be told that his study is a fundamental one. This truth, however, has not yet fully penetrated the world at large"[1].

The possession of a pedigree was for generations a sign of noble birth, distinguished ancestry, and privileged position. The problem for the family historian today is that the snobbery associated with pedigrees, the way in which they have been manipulated and the uses to which they have sometimes been put, the notorious ingenuity of genealogists, let alone the immemorial vice of making false pedigrees have all combined to bring the subject into disrepute. The most ancient manuscript pedigree and the most recent multi-media family history on compact disk downloaded from the Internet may be equally worthless without that which is so often lacking: exact knowledge of the sources and methods used at every stage of their compilation.

Yet we all know that the history of individual families when put beside the history of other individual families begins to form the history of a locality, a community or a social group, and that when all these are put together we have a large part of "history" itself.

Almost anyone may aspire to be a skilled genealogist in the sense that he or she can draw a pedigree but he or she may be quite uncritical of the pedigree so constructed. The early Elizabethan heralds were the first to use the drop-line chart pedigree, so useful for quickly showing the relationships of a family group, but they have an especially bad name for the making of false pedigrees. A critical science of genealogy did not, in any case, then exist though it soon began to develop.

The exploration of documentary sources by which to establish and prove the relationships set out in a pedigree, in earlier times usually for legal purposes (when the names of wives, unless they were heiresses, were of no importance), the later development of the writing of family history, and more recently the relating of that history to the social and economic history of the time, has been a slow process over a long period.

After the Norman conquest it was that legal aspect of individual pedigrees for particular purposes, as in the lengthy statements of descents on the Plea Rolls, which predominated for centuries. Most of these statements were probably based on orally transmitted knowledge, though some of the longer ones may have been compiled from written sources, as in the Scrope versus Grosvenor case of 1378 when charters were produced in evidence.

It was not until the fifteenth century with the development of other antiquarian and topographical studies that collections of pedigrees were made. The oldest books date from about 1480.

The rise in the sixteenth century of many new families to wealth and station in a society where the prestige of ancient blood was great produced, as it did in the nineteenth century, a market for deplorable concoctions as well as for genuine research. The series of heralds' visitations made in the 1560s recorded many lengthy but doubtful pedigrees as well as some fabrications, and it was not until those made in the 1580s that Robert Glover began to illustrate the principle that pedigrees should, if possible, be founded on record evidence. A working knowledge of the public records was first brought to the subject by Augustine Vincent who died in 1626. He was a pupil of William Camden and had worked at the Tower Record Office. An apprenticeship system has always been important in the practices of professional genealogists who thus had many advantages over the amateur working alone.

One of the first family histories to be compiled seems to have been that of the Berkeley family in Gloucestershire by their steward, John Smyth of Nibley, who died in 1640, using public records as well as the family's papers and charters at Berkeley Castle. The first to be published, replete with forged charters and fictitious pedigrees, however, appeared in 1685.

Sir William Dugdale showed his superiority in the field in the skill with which he marshalled the evidence in his *Antiquities of Warwickshire* (1656) and *Baronage of England* (1675-6), citing contemporary record evidence for every statement made. When he was deceived by spurious documents, as with those on which rested the claim of the Feildings, Earls of Denbigh, to descend from thirteenth century Hapsburgs, one knows exactly what these were.

The pedigrees of knights compiled by Peter le Neve show that by the end of the seventeenth century the pedigrees of newcomers to this class needed something more than a knowledge of the records of land tenure, and he began to use parish registers. Sir Comport Fitch, whose father had been a carpenter, registered a pedigree at the College of Arms in 1699 which had apparently been worked out for him by Samuel Stebbing. Stebbing had set about it by copying wills, making extracts from parish registers, noting monumental inscriptions, interviewing members of the family, and fitting all this evidence together as best he could. It is an early and elementary instance of what has since become a commonplace of genealogical method. Church monuments had always been used but half a century later the importance of those in the churchyard for humbler families came also to be recognised.

The peerage writers of the eighteenth century, Arthur Collins and Joseph Edmondson, almost exhausted that side of the market but the local historians of the eighteenth and early nineteenth century now poured out county histories containing pedigrees of the landed gentry, many embodying fabulous material as the price of a subscription, for as Thomas Whitaker wrote, "in the genealogies of old families there are many vestiges of error, and some of fraud, which time and vanity have rendered sacred".

However, the movement at the same time toward the publication and arrangement of the public records had a beneficial influence on the study and literature of the subject. The peerage claims of the 1830s produced a flurry of genealogical activity, much of it fine work, influenced by the publication of the first textbook, Stacey Grimaldi's *Origines Genealogicae; or The Sources Whence English Genealogies May Be Traced*, in 1828.

The nineteenth century with its great growth in the middle class saw the development of several large commercial genealogical practices. The Burke family, of which Sir Bernard Burke who died in 1892 was the most active member, conducted a considerable practice in research and publishing and is well known for the production of Burke's *Peerage* (from 1826), the *Commoners* and *Landed Gentry* (from 1837), the *General Armory* (1842-84), and numerous other works. The *Peerage* was based on Brydges' edition of Collins' *Peerage* and not only accepted the myths which Collins accepted but added more from information supplied by the families and from other sources. The newly compiled *Landed Gentry* contained even more unreliable matter. All were heavily embossed with the Royal crown and his badge as Ulster King of Arms.

E A Freeman, the historian of the Norman Conquest, fiercely attacked the "monstrous fictions" to be found in Burke's quasi-official compilations, but it was the critical genealogist Horace Round who, from about 1893 onwards, by his attacks on these fabulous pedigrees, as Anthony Wagner said, "preserved like flies in amber the follies and errors which he chose to castigate".

The vehicles used by these critics, which were a potent influence in raising standards, were a group of periodicals which published compiled genealogies and source material as well as reviews. One of them, Oswald Barron's *The ancestor* (1902-5), was widely read, not only for its brilliant scholarship but also for the elegant style in which he and Round destroyed the bogus descents, "nailing them one by one", as Round said, "as a gamekeeper nails his vermin". Thereby revealing the pleasures of destructive as well as of constructive genealogy!

Of the family histories published in the nineteenth century few have much merit. Among the best are R E Chester Waters' *Genealogical memoirs of the extinct family of Chester of Chicheley* (1878), Falconer Madan's *The Gresleys of Drakelowe* (1899) and A L Reade's *The Reades of Blackwood Hill* (1906). Some of the best researched chart pedigrees are to be found in the thirty-five volumes published by J J Howard and F A Crisp under the title *Visitation of England and Wales* which indicate the sources used and mirror, in their layout, pedigrees registered at the College of Arms. These give so much detail, however, that the relationships are often difficult to unravel and the basic purpose of an outline chart is defeated.

The rise of the Victorian middle class and the demand for professional genealogists produced Charles Bernau and George Sherwood. Bernau published in 1907 the first lists of families being traced, and George Sherwood was probably the first person to do a "one-name" search at the General Register Office in order to fix the distribution of the surname Boddington. Together they were instrumental, with a group of other professional people, in founding the Society of Genealogists in 1911, "to promote, encourage and foster the study, science and knowledge of genealogy by all lawful means".

One of the Society's aims was to bring together the various indexes and transcripts which were then being made and the collection of printed, manuscript and typescript copies of parish registers became a principal objective.

Just before its foundation the local historian W P W Phillimore, believing that "one of the chief obstacles to the completion of a pedigree, is the difficulty of obtaining the names of the wives", produced the first of 233 volumes containing transcripts of the marriage registers of about 1,650 parishes. The Harleian Society had also printed many London registers and half a dozen county societies and a general Parish Register Society were all publishing transcripts, mostly from their commencement in the sixteenth century to 1812 or 1837. The first county to be covered entirely in print was Bedfordshire, completed only in 1992. Progress was indeed slow but the Society's work was given urgency by the Second World War and impetus through its Committee for Microfilming Parish Registers which was active from 1939 to 1952. Using the transcribed registers then available Percival Boyd organised one of the genealogist's greatest working tools, a consolidated typescript index by county to marriages in England 1538-1837. It contains about seven million entries, perhaps thirteen per cent of the marriages which took place in that period.

Until at least the 1950s there was still a considerable element of snobbery in much of the genealogical work that was done, with an accent on "good" lines and royal descents. In general no necessity was seen to link the history of the families researched to the histories of the times in which they lived.

However, what was probably the first competent history of a yeoman family, William Miller Higgs's *A history of the Higges or Higgs family*, was published in 1933.

New records were being explored and following the great social upheavals in the first half of this century more people began to take an interest in the subject. The first of the new generation of basic guides was Arthur Willis's *Genealogy for Beginners*, published in 1955.

In 1961 the Society of Genealogists staged a small exhibition, which I organised, showing how one could trace the ancestry of a farm labourer. The position at that time was surveyed in the writings of Anthony Wagner, Garter King of Arms. His *English Genealogy*[2], published the previous year, had reflected the basic desire of the genealogist to extend his or her pedigree even further into the past, something taken to its limits in his *Pedigree and Progress*, with its tentative lines into antiquity, which he produced in 1975.

Pedigree and Progress came at the time of an explosion of interest in genealogy. Every type of ancestor, good or bad, came to be pursued, a process no doubt assisted by a generally more relaxed attitude to illegitimacy, lack of marriage, and the breakdown of family life itself. Indeed an inverted snobbery of descent from convicts became fashionable.

The catalyst seems to have been the publication and filming of Alex Haley's *Roots*, first published in Great Britain in 1977. In 1978 the Parochial Registers and Records Measure brought the majority of parish records into county record offices where they could be consulted without charge. The following year a popular BBC television series "Family History" showed everyone where and how to start. Meanwhile the Children Act of 1975 had given adopted people the right to obtain copies of their original birth registrations, thus opening the way for them to trace their natural parents and further ancestry for the first time.

The Society of Genealogists, today with fourteen thousand members, had remained the only society catering for the subject in England until one was founded at Birmingham in 1963. In the 1970s societies were founded in every county (there are now over a hundred of them), many in time acquiring numerous branches. Hampshire has as many as sixteen. Their members began to transcribe, index and publish local records and inscriptions in an unprecedented way. Together these societies probably have between eighty and a hundred thousand members.

A Federation of Family History Societies, representing their interests, was founded in 1974, and has itself done much to publicise and assist the subject through its twice-yearly conferences and a notable series of publications. The varied interests and objectives of family historians have also since 1984 been reflected in the pages of the popular monthly magazine *Family Tree* with its circulation of about forty thousand.

These co-operative movements in family history have taken place all over the English speaking world. Two members of the Society of Australian Genealogists have produced annually for the last seventeen years a volume called the *Genealogical Research Directory*, each edition of which now lists more than a hundred and fifty thousand families which people are tracing. Similar directories of "interests" are published by many local societies and by the Federation of Family History Societies, the latter with its three hundred thousand entries being called the *British Isles Genealogical Register* or "BIG-R". There is so much activity that it is difficult to find a family which is virgin territory where research is concerned.

Following the 1978 Measure the majority of parish registers over a hundred years old were deposited in county record offices where they have been the subject of much transcription work by the local family history societies. In two or three counties all the registers have been copied and several societies have compiled county marriage indexes, supplementing Boyd's Marriage Index. About half the counties are now involved in a major project to compile a National Burial Index.

An attempt to survey all the available transcripts of registers in public collections was made by the Society of Genealogists in 1939 and again in a series of volumes under the title *National Index of Parish Registers* from 1966 onwards, though so far the details for only about half the English counties and Wales have been published. Regularly revised catalogues of the copies in its own possession are published every two or three years. These now relate to some parts of the registers of over eight thousand, or about two thirds, of the parishes in England and Wales.

Just before the Second World War the Genealogical Society of the Church of Jesus Christ of Latter-Day Saints (the Mormons) commenced microfilming records and after the War started a programme to microfilm parish registers in the British Isles. Using computers and volunteer church members in America it has compiled from these microfilms the second of our major tools, the International Genealogical Index. First made available in England on microfiche in 1977, and now on compact disk, the Index, which is generally limited to baptisms and marriages in the period 1538-1875, contains about eighty million entries for the British Isles and about two hundred and forty million entries world-wide. All the existing baptismal and marriage registers for Scotland have also been indexed.

The part played by the Genealogical Society of Utah in the development of genealogy in the British Isles should not be underestimated. It has microfilmed vast numbers of records worldwide and many of these are now available through its ninety Family History Centres in England. In 1985 the British Genealogical Record Users Committee was organised by the Society of Genealogists to support that microfilming programme and it has since become an active forum for organisations representing users and keepers of archives.

The release of the 1841 and 1851 census returns to public search just after the Society of Genealogists was founded gave an impetus to the searches of those who knew little about their immediate ancestry. In the 1920s A T Butler at the College of Arms and others had begun to collect trade and commercial directories as a ready means of identifying the appropriate streets and houses in which the families sought might be found. Little systematic indexing by surname took place until the 1970s since when many surname indexes, particularly to the 1851 returns, have been compiled. The decision, taken by the Genealogical Society of Utah in association with genealogists in Britain in 1988, to transcribe and index the whole of the 1881 returns for England, Wales and Scotland, has revolutionised much nineteenth century research and again given great impetus to genealogists just commencing their search.

The importance of tombstone inscriptions as a source for those below gentry status had been recognised by Ralph Bigland in the eighteenth century. He was also the first person to copy them systematically, as his history of Gloucestershire shows. In this he was followed by numerous antiquaries and local historians and by the beginning of this century several genealogists were copying all the stones in their areas, the work of W B Gerish, who completed Hertfordshire, being a notable example. In the 1970s, and with many graveyards and cemeteries being destroyed to make space for car parks and because of the cost of upkeep, national coverage became an urgent object of the local societies, transcription parties were organised, and several counties have been completely or almost completely covered, those copied at the beginning of the century now needing to be done again.

Probate records have long been recognised as a major source of genealogical information and although some calendars of testators have been published by local record societies most work of this nature has been carried out by genealogists working in the British Record Society. The process was made much easier with local assistance following the transfer of the early probate records from probate registries to county record offices in the 1950s though funds for publication remain obstinately difficult to find. Again the work of the Genealogical Society of Utah has been important in this field. They having filmed practically all the probate records in the British Isles from the fourteenth century to the 1950s, the Society of Genealogists obtained from them copies of all the available indexes to the local courts and now has a major centralised collection. One of the Society's publishing achievements has been the production of six volumes of will indexes for the period 1750-1800 from the records of the Prerogative Court of Canterbury. The publication of full extracts from the wills themselves has proceeded much more slowly and a great deal remains to be done in this field.

I have touched on a few groups of records to illustrate progress in recent years. The Society's Library guide and the list of its publications will quickly show the value of the library and bookshop to anyone doing genealogical or biographical work. The resulting collections I believe, however, have an importance far beyond that of genealogy and family history.

A few genealogists in the past have been sufficiently inspired to attempt to trace everyone with their particular surname. With the growth of the subject such "one-name studies" have become very popular. The large bodies of information being collected, however truly blinkered the vision of the collector may be, are already showing the wider, demographic, uses of genealogy and surname distribution. Several associations of people with the same surname - one-name societies - have also been formed and a Guild of One-Name Studies was established in 1979 with regular meetings, a journal and an annual list of its members.

The transcription, calendaring and indexing of documents on a large scale by local family history societies and by some individuals and the collection of large quantities of data by genealogists, particularly those interested in one-name studies, was greatly aided by the advent of home computers and has in turn fuelled an interest in computers and in computer programs specifically designed for recording pedigrees.

A computer group was formed within the Society of Genealogists and in 1982 the Society started a quarterly journal *Computers in Genealogy*, with David Hawgood as editor, to report progress in the application of computers in genealogy. The management and arrangement of genealogical information, whether by drop-line chart pedigree, indented narrative, printed card or form, punched card or computer, is undoubtedly one of the attractions of the subject and the two disciplines go well together. The speed with which information can be transmitted by computers has, however, inherent dangers for the genealogist. Without proper referencing incorrect or fraudulent information is easily propagated and a more strict attention to the quotation of sources at all times is now constantly urged. There is, of course, another and more beneficial side and the co-operative efforts of widely scattered computer owners have provided a major source of voluntary indexers. One excellent such project has been that organised by David Squire to index the calendars of marriage licences issued by the Vicar General and Faculty Offices from about 1700 to 1850 the first part of which is now complete and published on microfiche.

When T V H FitzHugh's *How to write a family history*[3] appeared in 1988, although a fine work for those whose ancestors came from the professional classes, it seemed already dated. The first history of a working class family to be published and receive academic acclaim was probably Peter Sanders' *The simple annals* (1989). In this, however, the exact detail of names, dates and places, beloved by genealogists, gave way to a detailed social history. The limitations of a published work could not provide both and I began to think that things might have gone too far.

Genealogy by now, and not without some argument, had begun to be called "family history". Genealogy is basically the technique of establishing from documentary evidence the relationships between people. Family history relates a grouping of people to their environmental and socio-economic surroundings. However, the introduction of the term probably helped to give the subject some academic respectability.

With the steady increase in the numbers of interested people and the demand for professional assistance the need for professional standards came to be recognised. An Institute of Heraldic and Genealogical Studies had been formed at Canterbury in 1961 but the process was given greater strength from 1968 onwards by the annual publication of a list of competent professionals by the Association of Genealogists and Record Agents formed that year. The subject, however, continues to produce numerous part-time searchers, many far from competent and impossible to police.

The Society of Genealogists through its journal *The Genealogists' Magazine*, founded in 1925, had done much to improve the knowledge of the subject's sources over the years. The Society's regular series of lectures, some of which, like that on the "Companions of the Conqueror" in 1932, had received widespread publicity, was later extended to courses, week-end courses and conferences, subjects ranging from the general to the highly specialised. By the early 1990s we were organising sixty or seventy events a year. These had included overseas lecture tours and, from 1992, an annual Family History Fair attended by four thousand people. Since the 1970s many classes for family historians have also been organised by the local societies in addition to their usual monthly series of lectures and research surgeries. Others have been provided by WEA and University Extra-Mural Departments. These have all assisted in the process of improving the standards of the work done. Extra-Mural courses leading to a Certificate and later a Diploma in "Genealogy and the history of the family", for which I am the External Assessor, had commenced at London University by 1987. The Open University followed with its honours course "Studying family and community history: 19th and 20th century", which is being taken by many hundreds of students throughout the UK, and about which Professor Ruth Finnegan said, "it is not a backward looking study, but a way of understanding what is going on in this country today". The course produced its four volumes of *Studying Family and Community History* (1994)[4] which have become standard text books for those who wish to put their family in some sort of historical context. The Open University has also broken new ground by publishing many of the projects undertaken by its students on compact disk. Because of their bulk it is undoubtedly in this form that many family histories, both of the traditional kind and as social histories, will survive in the future, if they survive at all.

The appetite for yet more basic guides to the subject continues and has recently been more than satisfied by the publication of Mark Herber's 675-page *Ancestral Trails: the Complete Guide to British Genealogy and Family History* (1997)[5] which, in June, was awarded the McColvin Medal for an outstanding reference work by the Library Association.

When the Public Records Act was passed in 1958 it embodied the principle that, in the words of the Grigg Committee (1954), "no attempt should be made to keep in the Public Record Office things which would not otherwise be preserved solely because they contained information which might be useful for genealogical or biographical purposes". However, a generation later the Wilson Committee's *Report on modern public records* (1981) noted that the older genealogical or antiquarian uses of records had been "absorbed into a richer and far wider study of family history, local history and military history" and concluded: "We consider that these widespread interests in the history of the nation, the family, locality and other groups ... are an important and wholly desirable development in national cultures".

Since then there has been a considerable shift in emphasis and some attention to the needs of genealogists when the preservation or destruction of records is being considered, but much remains to be done. The Society of Genealogists and the other societies through the Federation of Family History Societies have now a powerful lobby in the field which has been used to good effect in a number of cases.

However, there is one field in which we have not succeeded. The centralised civil registration of births, marriages and deaths was introduced in England and Wales in 1837. The original certificates and some other records dating back to 1761 remain closed to public access and information is available only on the payment of heavy fees for copies. That lack of access to the genealogist's basic records remains a major deterrent to many new researchers and an obstacle to a large number of other forms of research.

The present situation in the British Isles mirrors that in America where easy access to a large number of transcribed and indexed documents often means that only a limited knowledge of social history or palaeography is necessary for the construction of a pedigree. Its practical effect is that most work by the local societies in England concentrates on the modern period, and Americans coming here to search for ancestors in the seventeenth century find any work in original records doubly difficult.

As an interest in the subject has gone further down the social scale the exploration of more recent records of every description has grown apace. For many access to First World War soldiers' records is now of more vital importance than access to those of the eighteenth century or of earlier periods.

However, there is now amongst these genealogists, as Michael Erben of Southampton University noted, "a group of researchers with an astonishing grasp of a huge and daunting range of sources and displaying an historical sensitivity and acumen that would be the envy of some professional historians".

Modern family historians have shown in a dramatic way that the past does not belong only to the professional and to the scholar. The energy, industry and determination which they bring to their subject is proverbial. In their "unashamed glee" they enjoy gathering the fruits of the past. They have learned, with the local historian Reginald Hine, that "Not rough, nor barren, are the winding ways of hoar antiquity, but strown with flowers".

However, one has to recognise that many think of the subject merely as an amusing and absorbing hobby. Many have little interest in producing a documented family history and do not see their work as having any lasting value. The challenge, perhaps, is to preserve the best of their work for future generations.

References:

[1] Anthony R Wagner, *Pedigree and Progress: Essays in the Genealogical Interpretation of History* (Chichester: Phillimore & Co., 1975).

[2] Anthony R Wagner, *English Genealogy* (Chichester: Phillimore & Co., 1983).

[3] Terrick V H FitzHugh, *How to Write a Family History* (Sherborne: Alphabooks Ltd., 1988).

[4] Rosemary Finnegan, Michael Drake & W T R Pryce, eds., *Studying family and community history* (4 vols.; Milton Keynes: The Open University, 1994).

[5] Mark Herber, *Ancestral Trails: The Complete Guide to British Genealogy and Family History* (Thrupp: Sutton Publishing Limited/Society of Genealogists, 1997)

GENEALOGY

Bibliography:

Anthony J Camp, *Everyone has Roots* (London: W H Allen & Co. Ltd. and Baltimore: Genealogical Publishing Co., 1978).

Anthony J Camp, "Family History" in David Hey, *The Oxford Companion to Local and Family History* (Oxford: Oxford University Press, 1996), pp. 168-174.

John Titford, *Writing and Publishing Your Family History* (Newbury: Countryside Books/The Federation of Family History Societies (Publications) Ltd., 1996).

The Galton Lecture 1998: Eugenics: The Pedigree Years

Pauline M H Mazumdar

Eugenics – the "science of human betterment" – was a movement that attracted progressive minds around the world in the early decades of this century. To say that it had a lot of support is an understatement: it swept through middle class intellectual communities in Europe and the Americas, both North and South, where each country had its own version of it as propaganda, research and legislation. There was virtually no one who had anything critical to say about it until the late nineteen twenties.

The classical methodology of eugenics was the pedigree study. In this paper, I shall look at the way pedigrees were developed and used by the three leaders of the movement, Britain, the US and Germany. Each of these countries had its own version of the movement, each with its own sources, history and outcome, but all of them collected thousands of human pedigrees. In the early years, it was enough to demonstrate *ad oculos* that like produces like. But during the twenties all of them were faced with the problem of making the transition from the pedigree as rhetoric to the pedigree as an investigative tool. They had to try to find a way of analysing the material they had collected.

We are here today to commemorate Francis Galton, so let us begin with him, and with the British movement. Although Galton first appropriated the pedigree diagram to scientific use in 1869 it fell to others to develop it and he did not pursue it in his later works. His *English Men of Science* of 1874 and his *Noteworthy Families (Modern Science)* of 1906 are collected

EUGENICS: THE PEDIGREE YEARS 19

from the rolls of the Royal Society. He describes the individuals and their families in words, rather as contemporary works described famous thoroughbreds and their relationship to the original Arabian foundation stock, together with their racing successes.[1]

Figure 1: from Thos H. Taunton, *Portraits of Celebrated Racehorses of the Past and Present Centuries, in Strictly Chronological Order, Commencing in 1702 and Ending in 1870* 4.v (Low, Marston, Searle & Rivington, 1887) v.1, opp. p. 16)

This is Flying Childers, famous for having a speed of a mile a minute, foaled in 1715, by the Darley Arabian, out of Betty Leedes, by Old Careless, out of a sister to Leedes, by the Leedes Arabian. Some of these horses never raced, like the female kin of Galton's FRSs, but they could still pass on great qualities.

The Eugenics Education Society, as this society then was, began to collect pedigrees in about 1910. The Society's Research Committee was set up in response to the Report of the Royal Commission on the Poor Law, with the mandate of looking into eugenic effects of the Poor Law. The Research Committee's work perfectly demonstrates the link between the pedigree method and the Society's ideology, its focus on the problem of the urban poor, called the "residuum." These were the people who worked in what is now the informal sector: hawkers, rag pickers, casual labourers, char-women, sex workers, people who at one time or another called on the support of the Poor Law, and so formed an administrative pauper class. Pauperism had been a problem that fascinated the activists of the Victorian middle class. Various different groups had suggested that it was caused by lack of sanitation, by lack of education, especially moral education, or by indiscriminate charity. The Society's Research Committee was bent on proving that it had a biological cause.[2]

To support their claim, the Research Committee set up an investigation of actual pauper families. The Relieving Officers of three workhouses co-operated with about twenty members of the Society to trace the family histories of the paupers who had been in the Workhouse. The Committee concluded that

> a single family stock produces paupers, feeble-minded, alcoholics and certain types of criminals. If an investigation could be carried out on a sufficiently large scale, we believe that the greater proportion of undesirables would be found connected together by a network of relationship ... [3]

The emphasis from the beginning was on this network of relationship. One of the three Relieving Officers who helped the Committee was E J Lidbetter, a man who had worked for the London Poor Law Authority since 1898, with the responsibility of investigating applicants for relief. He found his life's work in the Society's project: he became a member almost at once, went to the training courses the Society put on, and soon became a frequent speaker at its meetings. Lidbetter was to spend the rest of his life collecting pauper pedigrees.

EUGENICS: THE PEDIGREE YEARS

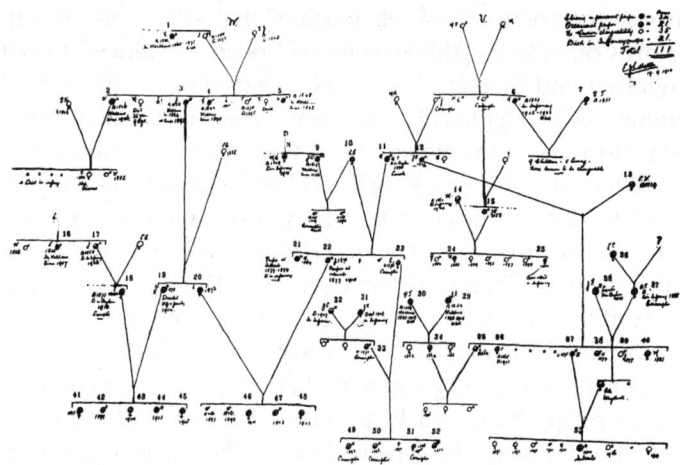

Figure 2: A Lidbetter pedigree, as published in his "Some examples of Poor Law eugenics," *Eugen. Rev.* (1910-1911) 2 pp.218.

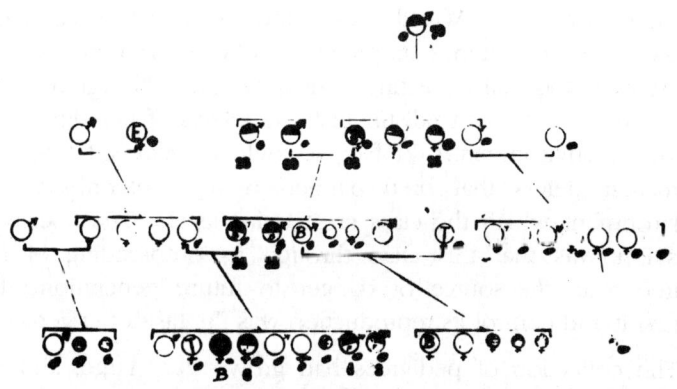

Figure 3: A Lidbetter pedigree, from a lantern slide at CMAC (?1912, for 1st Congress.)

It was the "network of relationship" that interested Lidbetter and his Committee. Although he had been introduced to both Mendelism and biometry as complementary ways of looking at patterns of inheritance, he was not concerned about mechanisms of transmission or genetic theory. He and his Committee saw the entire pauper class as a kind of breeding isolate, to use a later concept. They were interested as much in the intermarriages between families as they were in the transmission of discrete defective traits.[4] Their pedigrees look like sections of matted felt, criss-crossed with relationships. One of the speakers at last year's symposium referred to them as "Lidbetter's absurd pauper pedigrees," and marvelled that behavioural problems such as unemployment and pauperism should be put down to genetic defect rather than to economic factors.[5] In my view, this may not be present day opinion, but it was not an absurd aberration. It was the core of the Society's position and the foundation of the eugenics problematic that it was to represent for many years to come.

After the First World War, the Research Committee reassembled to continue its project. Lidbetter, with occasional assistance, was still collecting pedigrees, and still arguing that they were not meant to distinguish the effects of heredity from those of environment. In fact, it could be claimed that the particular defect that bred pauperism also determined the environment which the class created for itself. The Society's position was that this class through its outbreeding of its betters was *the* source of danger to future generations; to define it and control its reproduction was the task of eugenics.

The collection of pedigrees had grown very large, and in time, difficult to organise, even though the project was always short of funds and Lidbetter rarely had more than one helper. Lidbetter kept an index of the people whose names he knew, but it was not easy to get the material into a usable condition. His intention seems to have been to create a complete map of the pauper class and its network of relationships. He seems to have regarded the class as a whole as a biological isolate rather like the fauna of the Galapagos Islands. His difficulty in expressing this is mirrored in mine in getting his pedigrees on slides.

The value of the pedigrees was in their complexity: in their extending over several generations, and showing the family connections within the class. "Paupers," he wrote in a Memorandum to the Committee, "may be an isolated group, and pauperism not a sporadic characteristic of the population as a whole, but a differentiating character of a small but clearly defined group."[6]

In both US and Germany, pedigree collection was linked explicitly from the beginning with the argument for Mendelian transmission of traits. But if you look at the Lidbetter pedigrees, you will see that there is no argument here for Mendelian transmission, even after the conventions had been standardised on the US model. There is no attempt at quantification, no propositus or index case, no attempt to disentangle either the families or the different traits, no attempt to determine what was environmental and what was truly genetic, and no control group.

In fact, those were the very features that were criticised most strongly after the War, when the statistician R A Fisher joined the Research Committee. Fisher himself probably shared Lidbetter's point of view to some extent. He had written that a human population does not mate randomly: the feeble minded, the likely source of the pauper class, were not evenly distributed throughout the population; they might make up one sixteenth of an intermating group which itself might constitute 5% of the whole population. He may actually have got that idea from the Lidbetter study, but where Fisher speaks in percentages of population, Lidbetter simply wanted to collect the pedigrees of individuals and show their relationships, which made statistical analysis very difficult. The Committee tried without success to get funding for a control group, but it was difficult to know how to design one. When the Committee applied to the Medical Research Council for a grant, the response was to ask for an outline of the statistical treatment proposed. By 1923, the scientific world was not prepared to accept a study without a statistical analysis of some kind.[7] The study was finally published ten years later in 1933, with no statistics, only the promise that this was the first volume, and analyses would follow in a later one. It was too late: there were no more. The time for the simple pedigree as an investigative tool was over, even though it still had its uses as a very effective persuasive device. The Society still displayed them at exhibitions and other public meetings, and pointed out the moral and biological lessons they conveyed.

The eugenists in the United States also relied on the pedigree method, though in their case it always had a Mendelian implication. The American movement was an outgrowth of agriculture, not, as in Britain, of social activism. In 1899, American agriculturists came to London for the first International Conference on Hybridisation, sponsored by the Royal Horticultural Society. They came home inspired to found their own American Breeders Association, for the advancement of science in agriculture, of knowledge of the laws of heredity and of the practise of breeding. Its members were mainly connected with state colleges of agriculture, experiment stations or the US Department of Agriculture, people to whom genetic science promised to be the key to rapid improvement in plant and animal breeds.[8] Breeders and scientists quickly accepted the Mendelian picture of unit characters that passed unchanged from generation to generation. It had convincing practical value.

In 1906, the American Breeders Association formed a system of forty-three committees, divided into three areas of interest, plant breeding, animal breeding and so-called General Subjects. Most of the committees dealt with practical agricultural breeding methods for corn and wheat; cows, chickens and horses. Only four of the forty-three were focussed on research: they were Animal Hybridisation, Pedagogics of Breeding, Theoretical Research on Heredity, and Eugenics. In 1910, the Association started a journal, *The American Breeders Magazine,* and at the same time, reorganised itself into three sections, Plants, Animals and Eugenics. The leader in the eugenics area of interest was Charles B Davenport, founder and director of the Station for the Experimental Study of Evolution at Cold Spring Harbor from its beginning in 1903 until he retired in 1934.

The eugenics committee of the American Breeders Association was the first US eugenics organisation. According to Davenport, its duties might be summed up in three words – investigation, education, legislation. Subcommittees were to be struck to deal with the "protoplasmic basis" of various diseases, along with criminality and pauperism. Unlike its British counterpart it also included "mongrelisation" among present threats to the quality of population. But like the British society, it interested itself in the so-called feeble-minded. A Subcommittee on Feeblemindedness was designated to answer the question of whether two imbecile parents ever begat normal children. The presumed answer was already clear, however:

> ... As for the idiots, low imbeciles, incurable and dangerous criminals, they may under appropriate restrictions be prevented from procreation – either by segregation during the reproductive period or even by sterilisation. Society must protect itself; as it claims the right to deprive the murderer of his life, so also it may annihilate the hideous serpent of hopelessly vicious protoplasm. Here is where appropriate legislation will aid in eugenics and in creating a healthier, saner society in the future. [9]"

26   HUMAN PEDIGREE STUDIES

Annihilating the hideous serpent," of course called for the collection of pedigrees, for both investigation and persuasion, but mostly, as you can see from that quotation, to persuade legislators of what the investigators already knew to be true. Among the earliest moves was the standardising of the pedigree format and symbols, an improvement that was quickly taken up by Lidbetter's Research Committee in London.10

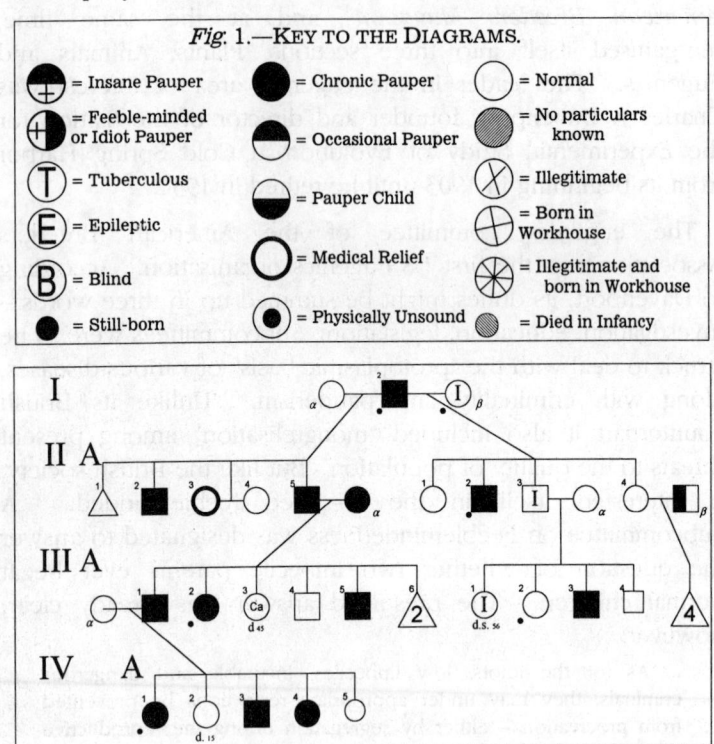

Fig. 1.—KEY TO THE DIAGRAMS.

- = Insane Pauper
- = Feeble-minded or Idiot Pauper
- = Tuberculous
- = Epileptic
- = Blind
- = Still-born
- = Chronic Pauper
- = Occasional Pauper
- = Pauper Child
- = Medical Relief
- = Physically Unsound
- = Normal
- = No particulars known
- = Illegitimate
- = Born in Workhouse
- = Illegitimate and born in Workhouse
- = Died in Infancy

Figures 5a and 5b: Standardisation – from Carr-Saunders, Greenwood, Lidbetter & Tredgold, "Standardisation of pedigrees – a recommendation," *Eugen. Rev.* **(1912-13) 4 383-390.**

Davenport's greatest success in the promotion of eugenics came through his contact with Mary Harriman, the rich and philanthropic widow of a railway magnate. According to

Garland Allen, Mary Harriman, like others at the time, had an interest in efficiency and scientific method; she tended to give funding to organisations devoted to helping individuals become more efficient members of society. She was attracted to Davenport's plans for solving social problems by the scientific study of hereditary social traits. Davenport wrote that she often said that the fact that she was brought up among well-bred racehorses helped her to appreciate the importance of a project to study heredity and good breeding in man.[11]

Mary Harriman's funding allowed Davenport to buy up the estate next to the Station for the Experimental Study of Evolution at Cold Spring Harbor, and use the house for a eugenics centre, to be called the Eugenics Record Office.

THE EUGENICS RECORD OFFICE
This is the hub about which the eugenics activities of the Carnegie Institution of Washington revolve. Here cooperation is offered to amateur family archivists or to research specialists on equal terms; here Dr. Harry H. Laughlin presides over the most extensive body of family trait records in America.

Figure 6: the Eugenics Record Office's house at Cold Spring Harbor, on Cape Cod; from *Eugenics* (1928) 1 15-19 (p.16.)

It was specially designed for the collection and storage of pedigrees, as well as for education about eugenics. It had a large main hall, and a storage room off it that had been lined with concrete and steel to make it fireproof. Harry H Laughlin, who had been teaching in the agriculture department of the State Normal School in Kirksville, Missouri, was appointed Superintendent. He was a devout eugenist, a true missionary, and also a man with a professional interest in breeding.

Laughlin began his work at the Eugenics Record Office in October 1910. He and Davenport had already trained a group of twelve field workers in pedigree-gathering. Laughlin describes the training and the principles behind it in his first report:

> ... the work included instruction in the principles of heredity, the distinction between worthless and telling records, and the practical methods of gathering significant data at first hand in the field. ... it is now generally recognised that the rapid advance in discovering the laws governing the inheritance of mental and physical traits is due to two things: First, the modern field method of getting at the inheritance distribution of biological facts, and, second, the modern analytical study of these facts, fitting them to the presence-and absence hypothesis and to the Mendelian principles.[12]

The plan is very similar to the Lidbetter project, with its focus on the pedigree of human failings, but here the emphasis is on fitting the transmission of failing traits into a Mendelian mould. Already in eight months of work, the six investigators chosen from among the first trainees to be in the house team had collected 626 pages of pedigrees and 2,270 pages of descriptions. They had worked on families with epileptics, criminal psychopaths, vagabonds, consanguineous marriages, albinos, and the feeble-minded. They had access under medical supervision, or were themselves doctors, to patients and records of the Skillman School for Epileptics, the Chicago

Psychopathic Institute, Matteawan State Asylum in New York State, and the New Jersey State Home for Feeble-Minded Women. The Record Office wrote round to heads of all institutions in the US dealing with abnormal individuals, and found, Laughlin states, that all supported the eugenic initiative.

This was a much bigger operation than the one in London. It had more money, since it was funded by the Carnegie Institution, and it was better organised and staffed. The Record Office had a newsletter, *Eugenical News,* at a subscription of 50c. a year, post free, subsidised no doubt. You can read in every issue how many more pedigrees had been accessed, how many new field workers had been trained – by the time of the 1921 class, a total of 215 persons had been through the course.[13] The newsletter gives details of the clinical experience offered the class of 1921. They went on a visit to Kings Park State Hospital, for a lecture-demonstration on types of insanity, as well as to the Brunswick Home for the Feeble-minded at Amityville, New York, and other homes for the feeble-minded, and to the New York State Hospital for Crippled Children and the Ellis Island Immigrant Station. There they had a demonstration of unsuitable immigrants detained and awaiting deportation. In the evening of that day, they went to Coney Island Amusement Park, and held "impromptu clinics" – I'm quoting the newsletter – at the side-show stalls of various human freaks, such as dwarfs, giants and microcephalic idiots.[14]

One finds this mixture of racism and prurience pretty rebarbative today, but at the time, it seemed to be a very impressive training for the investigators. It is actually quite difficult to find any publications making use of the Record Office collection, other than as an example of how to collect the material itself; but it was certainly displayed, much as the Eugenics Society's posters were, as a powerfully direct means of persuasion.

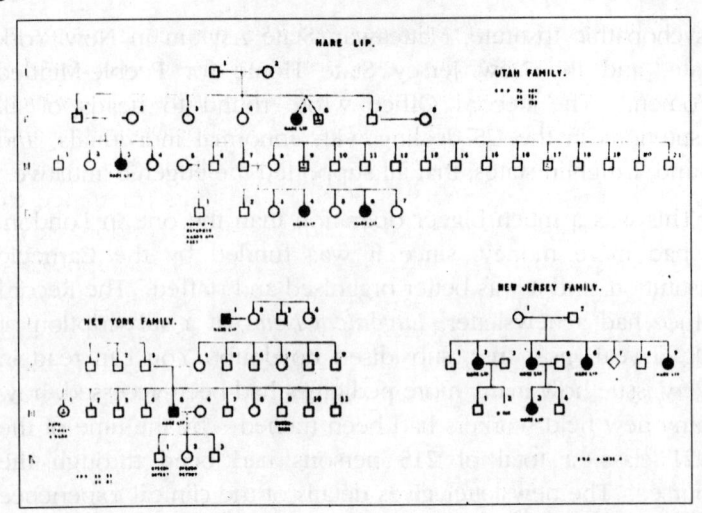

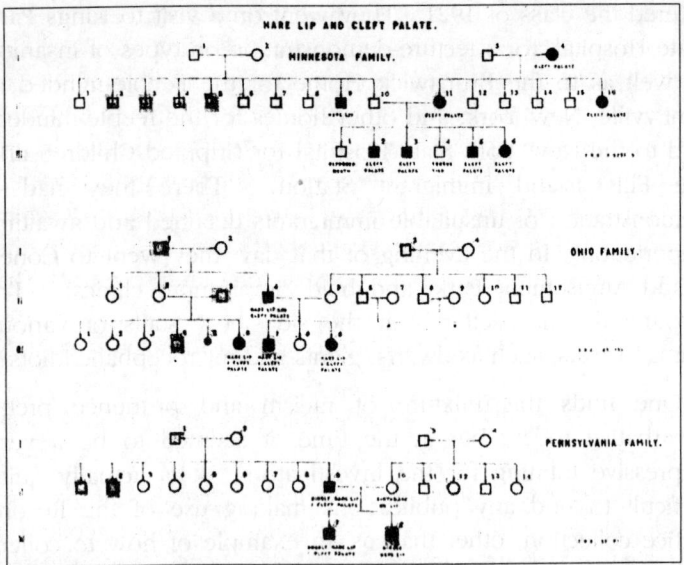

Figure 7: American pedigrees from Laughlin, ed., *Second International Exhibition of Eugenics*, 1921.

Yet, like the Lidbetter Pauper Pedigree project, it began to come in for criticism from a scientific point of view in the twenties. The Carnegie Institution of Washington, which supported both Davenport's Cold Spring Harbor laboratory and the Eugenics Record Office under Laughlin, was being pressured by critics. In 1928, it sent a visiting committee to inspect the work of the Records Office. As in London six or seven years earlier, a group who were themselves supporters of the eugenics movement produced a report that was very critical of the Record Office's methodology, especially of the failure to analyse the material. There had been no effort, they said, to develop a quantitative technique, and the majority of the records depended on the subjective assessment of an individual fieldworker. The material would have to be tested on a few genetical problems to see whether it was of any use. A second visiting committee in 1935 was much harsher: it rubbished the whole operation. The collected records were worthless for human genetics, it stated, and the collection should be stopped as each project came to an end.[15] The parallel with the Lidbetter project could not be closer, except that his had come to an end two years before, in 1933, with a courteous publication of some of his collection. Laughlin himself lasted a few more years, until his health gave the Carnegie Institution the opportunity to force him to retire. The Eugenics Record Office closed on New Year's Eve, 1939, and Laughlin, poor soul, left Cold Spring Harbor for ever in January 1940.

My last example of a pedigree culture in full bloom comes from Germany.

Like the Americans, the German eugenists took up Mendelism very quickly. But the pedigrees that they collected were subjected to analysis by the most sophisticated mathematical methods. These were introduced by Wilhelm Weinberg, a statistician and hygienist from Stuttgart, who was the founder of the Stuttgart branch of the Gesellschaft für

Rassenhygiene, the German eugenics group. In 1908, Weinberg was already generalising Mendel's original binomial formula.[16] In 1909, he devised a statistical test that would show whether a given trait was inherited as a Mendelian factor. If it was, the ratio of affected to unaffected in parents and children should differ from the same ratio in the siblings. If it did not, Mendelian inheritance must be ruled out. He also introduced corrections that would deal with the problem of excess affected children in a seemingly recessive situation – the problem here being that when phenotypically normal parents carried some recessive gene, their children would not be counted into the calculation of a Mendelian ratio if none of them were visibly affected. They would seem to be a normal family, and nobody would notice them. Conversely, when there were several affected children, the family would be seized upon by excited researchers and published with glee. These were important problems where the researcher had to pool data from several families to get around the sheer microscopic size of the human family, compared to, say, fruit fly families. Pooled data therefore always showed more than the proper Mendelian recessive ratio of 25% affected children. A well-known Swedish researcher, Hermann Lundborg, working on a huge family with 2,232 members, got 32% affected children for the supposed recessive trait he was studying. He sent the pedigrees to Weinberg, who worked out corrections that would take the missing phenotypically normal families into account. The collection was published in 1913.[17] You can see already that both Lundborg and Weinberg were into statistical analysis of their material. They were not content, – well, Lundborg might have been, but Weinberg put him right – they were not content, as the US and the British groups were, simply to exhibit a shocking pedigree for its rhetorical or promotional value. These workers were pooling data, testing for fit with expected Mendelian ratios, and dealing mathematically with the results. As R A Fisher pointed out to the Eugenics Society's Research Committee in 1923, you simply could not do this with Lidbetter's pedigrees, collected with nothing of the kind in mind. Nor was it attempted by Davenport and Laughlin at the Eugenics Record Office. In fact, Davenport once wrote that there was not much more to Mendel's law than ½ of ½ is ¼.[18] Weinberg thought it would do him good to work through some Weinbergian calculations. And although Laughlin was very keen on German eugenics, and published summaries of papers from the German journal *Archiv für Rassen und Gesellschaftsbiologie* in his *Eugenical News* right up into the Nazi period, it seems that mathematical Mendelism did not attract him. Weinberg's papers were very difficult to follow.

The German eugenists took up mathematical Mendelism with enthusiasm. The young psychiatrist Ernst Rüdin published a long paper explaining Mendelian inheritance to fellow psychiatrists, with a pedigree diagram showing how the recessive trait appears and disappears, missing two generations in his example. He cheats a little in giving the last generation exactly 25% affected children.[19]

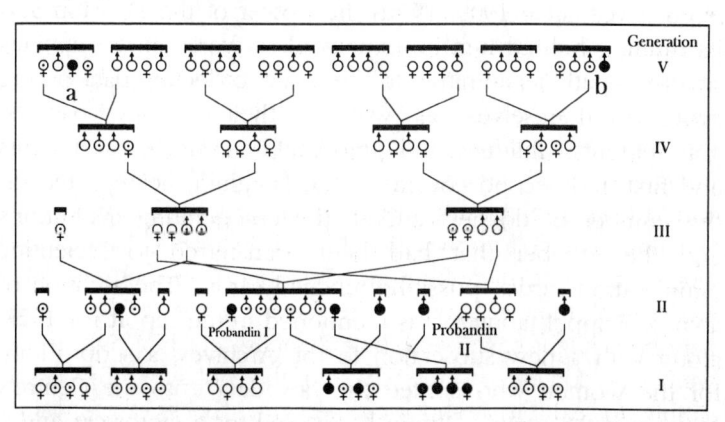

Figure 8: from Rüdin, "Wege und Ziele der Familenforschung mit Rücksicht auf die Psychiatrie," Z. f. d. ges. Neurol. u. Psychiatrie (1911) 7 487-585 (p. 498).

In 1916, he published a monograph on the inheritance of *Dementia praecox* or schizophrenia, in which he made use of Weinberg's methods. Rüdin took his cases from the pedigree collection of the Munich Psychiatric Clinic, which had a huge archive of family material. The director, Emil Kraepelin, was impressed. When a legacy from an American well-wisher allowed for the foundation of a Research Centre connected to the clinic, Kraepelin put Rüdin at the head of its Genealogy and Demography Division. The Centre nearly went under following the end of the First World War with the German currency collapse, but it was saved by the Rockefeller Foundation. The Foundation funded a large new building with an entire floor devoted to Rüdin's Division. Kraepelin died in 1926, just before the building was ready. When it opened in 1928, Rüdin was director in all but name.

The basis for the Genealogy Division's work lay in the clinic's huge collection of case material. Rüdin himself had been collecting pedigrees of schizophrenics and manic depressives since 1909. With the power of the Division and its funding behind him he expanded to work up the families of all the patients admitted to the clinic, collecting data on the probands themselves, as well as their siblings, parents, grandparents, children and grandchildren, uncles and aunts and first and second cousins. Like Laughlin, he reported on the number of dossiers added, the co-operating institutions and the number that had been transferred to "scientific enumeration cards," possibly punched cards. The addition of a "new Koppel tabulator" is mentioned in the report for 1928, along with a new subsection to the Archives, a scriptorium, for the women who copied the raw data onto the Centre's forms and the cards. Rüdin had asked for a motor car and a considerable travel budget, showing that the Division's staff spent a lot of time tracing the families of its patient probands. By 1929, the Division had collected more than 20,000 pedigrees. The following year, the Centre was trying to get a grant to buy from Weinberg a collection of pedigrees that he had put together from asylums around Stuttgart, with some of them going back as far as 1813. The cards from that collection had already been on loan to the Centre for some time. Here was pedigree-collecting on an industrial scale, which put even Laughlin's well organised Eugenic Record Office in the shade, to say nothing of Lidbetter's efforts.

Mathematical Mendelism was difficult. It was difficult to understand and to teach, and it was not something that could be explained to a potential political supporter. It did not have the appeal of the straightforward pedigree. Since the goal of a eugenics programme was not just research but teaching, legislation and action, it seems that with the Weinbergian methods of analysis Rüdin and his group had painted themselves into a corner. Beginning in the early twenties, Rüdin changed his position. He began to feel that the Weinbergian corrections were no more than an artificial attempt to force the figures into a Mendelian mould. The results were unreal, distorted by the things that Weinberg's corrections were designed to correct, and also by the difficulty of defining a disease trait, especially with schizophrenia. The Kraepelin school as a whole tended to emphasise that schizophrenia changes over time and from patient to patient and has no sharp boundaries. The patients' families all contained people who were not schizophrenic, properly speaking, but were still not normal. Rüdin decided to abandon Mendelism for what he called *empirische Erbprognose,* empirical genetic prognosis. The first paper to do this followed up 51 patients from the Munich clinic archives. Between them, they had 126 children, of whom 24 died too young to have shown symptoms. Of the 101 other children, only 37% were normal. The 63% abnormals were made up of 8% true schizophrenics, 48% that the clinic called schizoid psychopaths, and 5% otherwise abnormal, odd or eccentric. Two thirds of the pooled progeny of these parents was abnormal, a much more striking result than your corrected Mendelian 25%. Other members of the group took up nieces and nephews of patients, and their grandchildren.[20] Some collected a normal sample to compare with those from the clinic patients. A normal sample, of course, contained quite a few random eccentrics and dements. In this case, the normal controls were the spouses of the clinic's neuro-syphilis patients, and the spouse's families.

From these data, Rüdin and his colleague Hans Luxenburger produced what they called their "prognostic canon," giving the expected numbers of affected persons in the extended families of patients with schizophrenia, manic-depression, epilepsy and general paralysis of the insane, compared to the normal population. The very high gross percentage of abnormals given by this procedure was striking. Here were the kind of figures to make an impression on the legislators, and to magnify the importance of the work. By 1930, Rüdin was talking about the practical results of his method, and drawing up a sterilisation law.[21] In 1933, with the Nazi accession, he got it. It was the first of a series of Nazi laws that led up to the marriage laws of 1935, which covered both the health and the race of marriage partners. By 1935, the law had been widened to require sterilisation of the more distant relatives of a patient. The method of empirical prognosis, with its inflated numbers, formed a ready justification for it. As an argument for sterilisation, empirical prognosis with its high percentages of abnormals outside the probands' immediate family was well able to persuade legislators of the result of allowing psychotics to procreate unchecked. No doubt, it also laid a foundation for the euthanasia programme of the forties, when mental hospital patients of all kinds, children and adults, were put to death.

When the text of the 1933 legislation became known in the US, Laughlin greeted it in *Eugenical News* as a "model sterilisation law," and thought he saw in it a reflection of the draft law that he himself had provided to several American states. In 1929, Laughlin published a paper in the *Archiv für Rassen-und Gesellschaftsbiologie* on the development of the US sterilisation laws.[22] In 1936, Heidelberg University awarded him an honorary degree, something that in the political circumstances, probably contributed to his forced retirement.[23]

Rüdin was also in contact with the Eugenics Society in London, which was at that moment mounting a campaign to legalise sterilisation. Cora Hodson, the Society's Education Secretary, wrote to him in 1930, sending R A Fisher's pamphlet on the effect of sterilisation on the numbers of feeble-minded in each generation. The Society had used figures from pedigrees collected by the American group, she said, but American work was largely discounted in this country, and they would like to revise their pamphlet with his help. Rüdin replied with a five-page letter, detailing his methods and listing his unit's publications. The German researchers were more interested in psychosis than in feeble-mindedness and social failings, but he sent what he had. A further enquiry in 1932 elicited a ten page letter, in which Rüdin emphasised that we must compare the incidence of defect among patients' relatives with that in the general public. Only with control groups could one be sure how much defect was inherited.[24] He sent the Society a copy of a voluntary sterilisation law recently passed by the Prussian health ministry, a forerunner of the Nazi legislation that Rüdin most likely drafted.[25]

Stefan Kühl has noticed that these international contacts, including the institutionally structured ones such as the Eugenics Congresses of 1912, 1921 and 1932, and the many meetings of the Eugenics International, gave Rüdin's lawyer a means of defence during his post-World War II de-Nazification hearing. He was able to argue that this science was no more than the science of its day, supported by colleagues around the world.[26] The sterilisation laws in particular, were widely shared. As we saw a moment ago, Laughlin in the US thought he had designed the prototype of the German law – perhaps he had – and the British eugenists hoped to learn from Rüdin the trick of convincing a legislature.

It was clear by the mid-thirties that the simple pedigree methodology was not adequate for contemporary research. German mathematical Mendelism had been adopted by opposition groups both in US and in Britain, including many geneticists. The new mathematical methods contributed to the attack on the eugenics movement and what now seemed its naive approach, and the attacks gained strength in the thirties. It may have been the insistence on mathematisation and statistics that led the movement as a whole in the direction of demography and population studies. Ironically, the Germans themselves, after shrouding the pedigree in mathematics, found that it had lost its persuasive value. They had now stopped using mathematical models based even remotely on pedigrees, and were collecting pooled survey data instead, trying to build up an empirical basis for genetical prognosis for relatives of patients, and hence for their sterilisation.

But pedigrees had not disappeared. They had lost a good deal of their power as investigative tools, but they were still very effective as rhetorical devices. Throughout the thirties, the Society went on using them to convince the general public of the importance of heredity in human failings, and, less often, in success.

EUGENICS: THE PEDIGREE YEARS 39

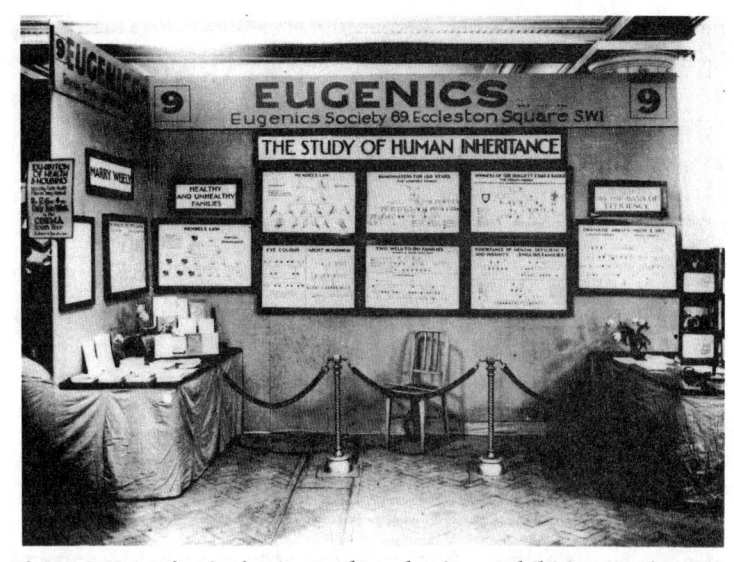

Figure 9: Eugenics Society's stand at a hygiene exhibition, in the mid-thirties.

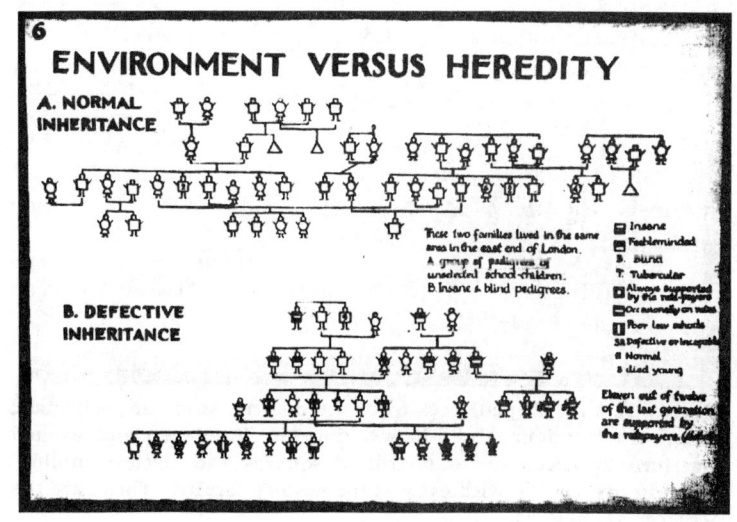

Figure 10: The control group: two pedigrees.

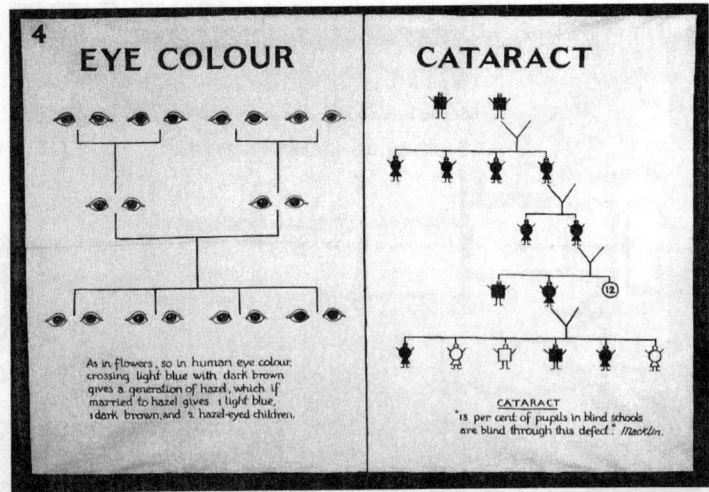

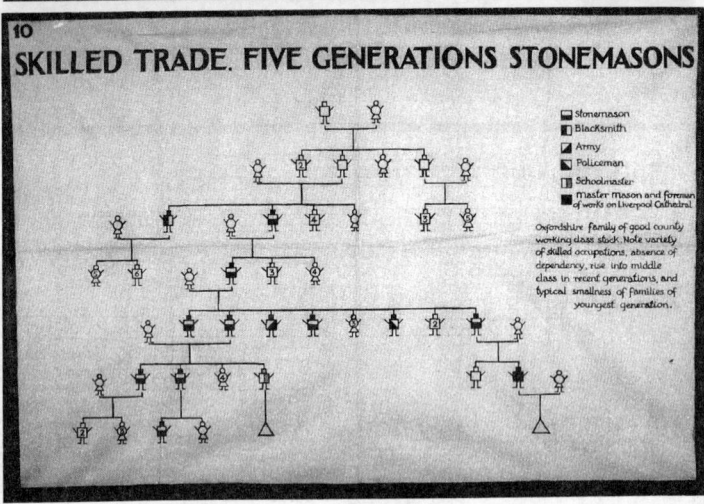

Figures 11 and 12: pedigrees from the display: some are schematic lessons in Mendelian inheritance; note the little men and women that have replaced the standardised squares and circles; another, showing eye colour with eyes; stonemason's family. These are the pedigree as rhetoric.

Although the Society had dropped the word "education" from its name in the early twenties, as it struggled with its research problems, education had never been dropped from its mission. The sterilisation campaign of the thirties, with its series of presentations to a variety of groups that might be harnessed to the legislative campaign, made pedigrees even more important for persuasive demonstration. They were now supplemented by Mendelian schemes for plant and animal species that lent the weight of well established scientific generalisation to the argument, and they tended to be clearer and simpler to understand, and better drawn, than Lidbetter's complex webs.

The pedigree methodology was a common factor in the work of all three of these leaders of the eugenic movement. That is no accident. The leaders met frequently through the international institutions of eugenics, the three large Congresses and the smaller eugenics internationals, that is, the Permanent International Eugenics Committee before the first World War, and the International Federation of Eugenics Organisations after it, where the leaders conferred more intimately as well as more frequently. They published in each other's journals, and each of the journals contains news of the doings of eugenics societies in other countries. They also had personal contacts and correspondence with each other. This was a truly international movement. So it is not surprising that developments in each country should more or less keep pace with the others.

The need to reach out to the educated public was also a common factor, especially with the visual material. Potential exhibitors for the 1921 Congress held in New York, with its accompanying popular exhibition and expected broad press coverage were told:

> While the exhibits must be able to withstand the test of professional scrutiny, still they should be of a nature which the man of ordinary intelligence and education, without special scientific training, may

readily comprehend and appreciate. Provision will be made for exhibiting displays of highly technical work, but the popular aspect will be given the preference.[27]

Number one on the list of exhibits that would be specially welcomed by the organisers was, "Human pedigrees which state the transmission of specific physical, mental and temperamental qualities." There were 22 other categories, but pedigrees came first.

I have argued that the core methodology of the eugenics movement, at least until the thirties, was the pedigree study. Beyond this period, even when its usefulness for investigation was under attack, it held its ground as a rhetorical device, a means of persuasion that successfully reached out to the general public. I rest my case.

References:

[1] Thomas Henry Taunton, *Portraits of Famous Racehorses of the Past and Present Centuries in Strictly Chronological Order, Commencing in 1702 and Ending in 1870, Together with their Respective Pedigrees and Performances Recorded in Full* 4 v. (London: Low, Marston, Searle & Rivington, 1887): see for example, Flying Childers (f. 1715), in v. 1, pp. 16-18.

[2] Pauline M H Mazumdar, "The eugenists and the residuum: the problem of the urban poor," *Bull. Hist. Med.* (1980) 54 204-215.

[3] "Report of the Committee appointed to consider the eugenic aspects of Poor Law Reform. Section I The eugenic principle in Poor Law administration," *Eugen. Rev.* (1910-1911) 2 167-177 (p. 173).

[4] E J Lidbetter, "Some examples of Poor Law eugenics," *Eugen. Rev.* (1910-1911) 2 204-228 (p. 225, Chart 2).

[5] Geoffrey R Searle, "Eugenics: the early years," in Robert A. Peel, ed., *Essays in the History of Eugenics* (London: Galton Institute, 1998) 20-35 (p. 26).

[6] Research Committee's Minute Book 1923-25, cited in Pauline M. H. Mazumdar, *Eugenics, Human Genetics and Human Failings: the Eugenics Society, its Sources and its Critics in Britain* (London: Routledge, 1992) 130.

[7] Mazumdar, *Eugenics, Human Genetics* (1992) N. 6, 131.

8. Barbara A Kimmelman, "The American Breeders Association: genetics and eugenics in an American context, 1903-1913," *Social Studies of Science* (1983) 13 163-204.

9. Charles B Davenport, "Report of Committee on Eugenics," *American Breeders Magazine* (1910) 1 126-129 (p. 129).

10. Charles B Davenport, "Conventional symbols for pedigree tables," *American Breeders Magazine* (1911) 2 73-74; A M Carr-Saunders, Major Greenwood, E.J. Lidbetter and A.F. Tredgold, "The standardisation of pedigrees: a recommendation," *Eugen. Rev.* (1912-13) 4 383-390.

11. Garland E Allan, "The Eugenics Record Office at Cold Spring Harbor, 1910-1940: an essay in institutional history," *Osiris* (1986) 2 225-264 (p. 236).

12. Harry H Laughlin, "Report on the organization and the first eight months work of the Eugenics Record Office," *American Breeders Magazine* (1911) 2 107-112 (p. 107).

13. "1921 Training Class for Field Workers," *Eugenical News* (1921) 6 61.

14. "Clinical and field studies of the 1921 Training Class," *Eugenical News* (1921) 6 61-62.

15. Allen, "Eugenics Record Office," (1986) N. 11, 250-251.

16. Wilhelm Weinberg, "Ueber Nachweis der Vererbung beim Menschen," *Jahreshefte des Vereines für Vater- u. Natur-kunde in Württemberg* (1908) 64 368-382.

17. Hermann Bernard Lundborg, *Medizinische-biologische Famienforschung innerhalb eines 2232-köpfigen Bauerngeschlechtes in Scweden (Provinz Bleckinge) Text und Atlas* (Jena: Fischer, 1913).

18. Cited in Charles E Rosenberg, "Charles Benedict Davenport and the irony of American eugenics," in his *No Other Gods: on Science and American Social Thought* (Baltimore, MD: Johns Hopkins, 1961) 89-97 (p. 91.)

19. Ernst Rüdin, "Einige Weg und Ziele der Familienforschung mit Rücksicht auf die Psychiatrie," *Z.. f. d. ges. Neurol. u. Psychiatrie* (1911) 7 487-585, p. 498, Fig. 7.

20. Pauline M H Mazumdar, "Two models for human genetics: blood grouping and psychiatry in Germany between the wars," *Bull. Hist. Med.* (1996) 70 609-657.

21. Gerhardt Baader, "Die Euthanasie im Dritten Reich," in G Baader and Ulrich Schulz, *Medizin und Nationalsozialismus: Tabuisierte*

Vergangenheit -- ungebrochene Tradition? (Berlin: Verlagsges. Gesundheit, 1983) 95-101.

[22] Harry H Laughlin, "Die Entwicklung der gesetzlichen rassenhygienischen Sterilisierung in der Vereinigten Staaten," *Archiv f. Rassen- u. Gesellschaftsbiologie* (1929) 21 253-262; cited in Allen, "The Eugenics Record Office," (1986) N. 11, p. 253.

[23] Allen, "Eugenics Record Office," (1986) N. 11, 253.

[24] Mazumdar, *Eugenics, Human Genetics* (1992) N. 6, 205-209.

[25] Mazumdar, "Two models for human genetics," (1996) N. 20, 653.

[26] Stefan Kühl, *Die International der Rassisten: Aufstieg und Niedergang der internationalen Bewegung für Eugenik und Rassenhygiene im 20.Jahrhundert.* (Frankfurt-am-Main: Campus Verlag, 1997.)

[27] Harry H Laughlin, *The Second International Exhibition of Eugenics, held September 22 to October 22 1921 in Connection with the Second International Congress of Eugenics in the American Museum of Natural History, New York* (Baltimore, MD: Williams, 1923) 16-17.

Human Pedigrees and Human Genetics

Elizabeth Thompson

Introduction

Genetics, the formal quantitative study of the inheritance of characteristics by offspring from their parents, began with Mendel (1866), although he studied peas, not humans. The earliest formal quantitative studies of the inheritance of human characteristics were 30 years later, by Karl Pearson and the early biometricians[1]. However, their studies of familial correlations in human pedigrees were not undertaken in a genetics framework. It was not until R A Fisher's famous paper[2], twenty years later again, that biometry and genetics became reconciled, with Fisher's demonstration of how similarities among relatives can be explained, using the principles of Mendelian genetics, with discrete genes segregating from parents to offspring, and different allelic forms of these genes having differential effects upon quantitative traits.

From 1890-1920, there was increasing interest in tracing traits in human families, and in a few cases, such as in the study of simple recessive diseases, these studies were placed in the context of Mendelian genetics. However, most of the increased genetic understanding from 1900 to 1930 came from studies of drosophila, mice, and other experimental organisms. By the 1920s, also, Mendelian genetic principles were applied to livestock in the early days of systematic selective breeding. In the English-language scientific literature, the analysis of genetic data on pedigrees was primarily confined to mice and livestock. However, in the early 1930s there is a sudden change, with Hogben's studies of segregation ratios in human families[3], and the recognition by J B S Haldane[4] and by R A Fisher[5] that the same ideas that had been applied to experimental organisms in the development of linkage analysis, and to agricultural plants and animals in analyses of quantitative traits, could be applied also to data on human families, or pedigrees. Since that time human geneticists have sought data on pedigrees, to test segregation ratios for diseases, to analyse similarities among relatives, to test for genetic linkage among traits, and more generally to resolve the genetic basis of traits.

Study of genetics and study of pedigrees go together, because genetics is about understanding the consequences of patterns of descent of genes in a pedigree. One can do genetics in a laboratory, but to relate the DNA sequences of genes to their phenotypic effects we need data on individuals. At the other end of the scale, one can study the genetic characteristics of populations, analysing patterns of differentiation and similarity at the population level. However, the population is a collection of individuals. It is at the individual and family level, in between the population and the test-tube, that the pedigree relationships among individuals affects the patterns of trait occurrences that we can observe.

Mendel's Laws and Chance Events

To return to Mendel (1866), in modern terminology his First Law states that at any given meiosis, at any given genetic locus, a random one of the two genes in the parent individual is copied to the offspring gamete. The gene which segregates in meiosis is independent of genes segregating in other meioses, from the same, or from other individuals. This law, which was stated by Mendel in almost exactly this form, is a probability law. It gives the chances that particular genes segregate down the lineages of a pedigree: genetically, a lineage is simply a sequence of meioses.

Chance events in the descent of genes within a specified human pedigree, is the focus of this lecture. Of course, there are also chance events in the formation of human pedigrees - who has offspring, which offspring survive, and in turn have offspring. It was these latter chances that Galton first focussed upon in considering the extinction of human surnames, or male lineages. However, Galton[6] was also aware of the contribution of chance events within a fixed pedigree, when he considered the average number of brothers of a male individual. In a family of fixed size, 2B, he does not have (B - 1) brothers and B sisters as an earlier note had suggested, but, on average, (B - ½) brothers and (B - ½) sisters. The difference results from chance variation within a pedigree (here a nuclear family), and the independence of meioses involving the segregation of the sex chromosomes from a father to his offspring.

Another example from the early history of Mendelian genetics is that of Fisher (1912), which Dr Anthony Edwards considered in his Galton Lecture in 1997. Fisher considered families in which the sons were equally disposed towards being "landed gentry" and to the "army", and discussed the number of sons the "landed gentry" son should have to compensate for the lack of family of the "army" son. Fisher inadvertently seems to have gained a factor of 2 here, probably because he considered only the male children. Let us repeat Fisher's example here, without prejudging the sex of the children, but, for convenience, using the male terms uncle and nephew. If an individual has two offspring, each of his genes will be represented, on average, once among those children. If an individual has four nephews, each of his genes will be represented, on average, once among the four nephews. Considering the grandparents, if an individual has four grandchildren, each of his genes will, on average, be represented once among those four grandchildren, and it does not matter whether these are four children of one child, or two children of each of two children, or any other combination.

However, there is more to this example when we consider the chance variation in Mendelian segregation. The expected number of gene copies is a statement of average genes shared, not of distinct genes represented. If an individual has two offspring, there is only chance 1/2 that both his genes at a locus will be represented among them. Even with four offspring, there is only chance 7/8 that both his genes are represented. The man with four nephews loses out to the man with two children. There is only a chance 7/32 that both his genes are represented among them. For the grandparents the difference is more severe. If each of two children has two children, there is probability 9/32 that both genes are present among the grandchildren. If only one child provides the grandchildren, at most one gene of a grandparent can ever be present among them.

This example becomes more complex yet, if we consider genes at linked loci, or an entire genome. In two children, or even four, the chance that one's entire diploid genome will be represented somewhere among them is close to zero. One needs to have about 10 children, before there is reasonable probability of transmitting one's entire genome[7]. Overall we see that relatively few genes survive, even over a couple of generations in a well ordered family. But expectations of gene copy number are preserved; the genes that do survive usually have several replicate copies.

Founder Effect in Human Populations

Considering again male lineages, and going back to Galton, suppose each male has an average of one male offspring. Provided each male has exactly one son, there is no variance in the process the lineages continue. However, as Galton was well aware, if there is any variance in the process, a mean of one leads to certain extinction of the process. If everyone has exactly two offspring, variance and random genetic drift are minimised, and effective population size is maximised at twice the population size. If the mean and variance of family size are each two, as, for example in a classic Poisson assumption, effective size is equal to actual size. If one half of the population has four offspring, and the other half none, family size variance is 4, and effective size is only 2/3 of the actual size. The greater individual variation in family size, the greater the potential for a few genes to become replicated in large numbers, while others become extinct. If family size is correlated over the generations, the potential is greater yet. The variance in pedigree structure, and the variance of Mendelian segregation within a pedigree structure, both contribute to founder effect.

In studying genetic isolates, it is often noted that certain traits achieve unusually high frequencies. Usually it is unclear to what extent these frequencies were truly extreme, and to what extent this is a result of ascertainment - the particular diseases studied in particular human populations are those which have achieved high frequencies in those populations. In any event, "founder effect" is often invoked as an explanation for the high frequency of an allele in a population. However, founder effect is not a cause, in the sense of a directional force such as selection. It is just an inevitable outcome of chance events in the descent of genes in pedigrees.

Consider, as did Fisher[8], a new mutation, or any single gene existing at some point in time. Suppose the population is increasing, so that, on average, each gene produces 1.1 genes, a 10% increase in the population each generation. Suppose the variance of the number of copies at the next generation is about 2. Then, assuming a particular form of offspring distribution which is probably a reasonable approximation, and absence of intra-generational correlations in family size, which probably is not, the probability the gene is extinct within 10 generations is about 85%, and within 20 is 90%. Conditional upon non-extinction, however, the expected number of copies of the original gene at 10 and at 20 generations are 17 and 67, respectively. Here, the probabilities combine the two sources of randomness introduced earlier - the randomness of the pedigree, and the randomness of descent of genes within the pedigree. The two sources contribute about equally to the overall variation in the numerical example considered here.

Few genes produce direct descendants, but the ones that do produce many. Moreover, the ones which survive will be the ones which, by chance, produced large numbers in the first few generations. We analyse the genes that are present, often because they are present in unexpectedly large numbers, in populations today. These will normally be genes with an *a priori* abnormal demographic history, ones that by chance had rapid initial increase in numbers.

To take a real example from my own past work[9], I was involved in a study at the Memorial University of Newfoundland, under the direction of Dr W H Marshall, of about 20 cases of immunodeficiencies and lymphatic cancers in three small West-Coast Newfoundland communities over the 20-year period from 1955-1975. There, the population of about 5000 had many founding ancestors, but 85% had some descent from one founder couple, who, with their children and grandchildren, first settled the area around 1820. Moreover, given the pedigree structure, many current individuals receive a substantial proportion of their genes from the original couple. We believed (I still believe) there is strong evidence for a recessive gene conferring susceptibility towards these immune-system diseases. We never found the gene; linkage analysis was never successful, but the pattern of inheritance was so clear one could even infer the ancestral paths of the allele. Researchers, and individuals of the communities, constantly raised the questions: Why now? Why so many?

Consider a single gene in one member of the original founding couple. They had 10 children and 70 grandchildren, which is a good start for any gene, but only 8 children and 24 grandchildren make any substantial contribution to the current population. The *a posteriori* inferred inheritance pattern indicates 4 children and 6 grandchildren who likely carried this putative allele. Moreover, the incidence, in timing and in number, is exactly what would be expected given that early history. Given these 4 children and 6 grandchildren as carriers (and likely others also with few if any current descendants), this chance early history accounts for the pattern of disease occurrence in the population. It is the chance early history that led to the multiple cases, and, if it had not, we would not be studying the disease in this population. Of course, chance and even genes are not the whole story. There may well have been some viral or environmental agent triggering the "epidemic" of cases in the years 1955-75, and susceptibility is only susceptibility, but it explains much to consider the pedigree probabilities.

Pedigree Relationships and Gene Identity Probabilities

When genes segregate from parent to offspring, they are normally an identical copy of the parental gene. Thus the gene is of the same allelic type as the parental gene. Although mutations may occur, for most genes the probability of mutation is small on the scale of a pedigree. More generally, genes which are copies of a single ancestral gene are called identical by descent (IBD), and, ignoring mutation, IBD genes are of the same allelic type. Our biological relatives are individuals with whom we share common ancestors, within some defined pedigree structure. Thus, they are individuals with whom we share, with certain probabilities, IBD genes. These genes must be of the same allelic type, and so have the same genetic effects, whether the gene is a blood type allele, causes a disease predisposition, or contributes to a quantitative trait such as a height or a lipid level. We have phenotypic similarities with our relatives because of these IBD genes; patterns of gene IBD underlie patterns of phenotypic similarity among relatives. The probabilities of gene IBD arising from a pedigree structure are the basic framework underlying analysis of genetic data on human pedigrees.

So let us consider some of these probabilities. Note again that they are probabilities; different realisations will vary. Suppose, for simplicity, the two parents of an individual are not related (and again, remember relationships are defined relative to a given pedigree). When we say we share half our genes with a parent, we mean precisely that. One half of our genome is a copy of one half of the genome of each of our parents. Or, for a single genetic locus, we share exactly one of the two genes at each locus with each parent. We also share, on average, one half of our genes IBD with our full sib, but, unlike the parent case, this is only an expectation or average over pairs of full sibs. At any given genetic locus we share both our maternal and paternal gene with probability 1/4, and share neither the maternal nor paternal gene also with probability 1/4, and we share exactly one of the two genes at the locus with probability 1/2. At the basic gene level, we are all equally similar to our parents, but siblings may share more or less of their genomes, depending on the chance events of meiosis.

So now let us take this chance variation in meiosis a stage further. Consider a grandparent and grandchild. The intervening parent shares exactly one gene with the grandparent, and exactly one with the grandchild; the probability that grandparent and grandchild share one gene at a locus is 1/2. On average, grandparent and grandchild share 25% of their genome, but the standard deviation is about 5%. Some grandparent-grandchild pairs may share as much as 35% of their genomes, others as little as 15%. Now first cousins share a gene at a given locus with probability 1/4, so for a pair of double-first-cousins (figure 1) the probabilities of sharing 2, 1, 0 genes IBD are 1/16, 3/8, 9/16 respectively. Thus these relatives also share 25% of their genome - the relationship is the same in terms of average IBD as that of grandparent-grandchild. But the probabilities are different, and hence patterns of trait similarity are different, even for simple Mendelian traits. For example, for a very rare recessive trait, the probability that given the grandparent has the trait, the grandchild will share it is small, of order the population allele frequency. However, given that an individual has the trait, the probability his DFC will share it is at least 1/16.

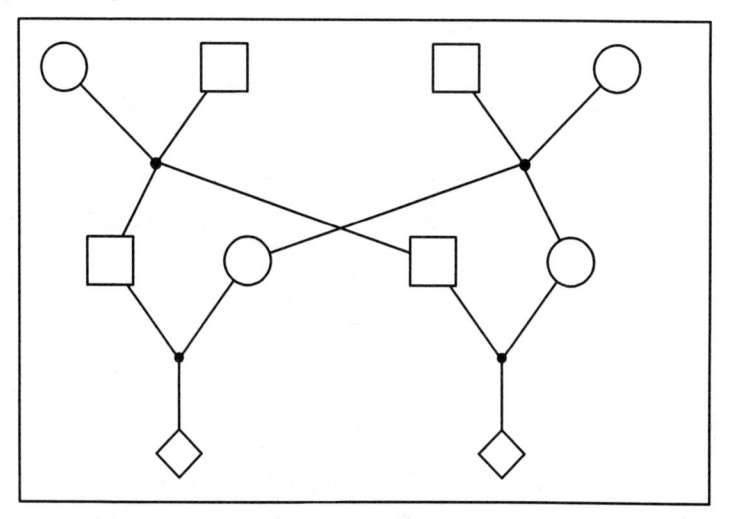

Figure 1: Double First Cousins

Other relationships also share this degree of relatedness, or average proportion of genome shared. One such is quadruple half first cousins (QHFC), although I admit to never having seen QHFC in a human population. This relationship arises when two couples each have a child, of opposite sexes, and these, in turn, produce a child. Meantime, the couples switch partners, and the reformed couples each produces a child, of opposite sexes, and these, in turn, produce a child. The two third-generation individuals are QHFC (figure 2). Each of the mother and the father of one individual is a half-sib to both the mother and the father of the other. However, the mother of each individual is not related to the father of that same individual; the individuals are not inbred. Without going in to the details of computation, the QHFCs share 2, 1, 0 genes at a locus with probabilities 1/32, 7/16, 17/32. Again the individuals share 25% of their genome, on average, or in expectation.

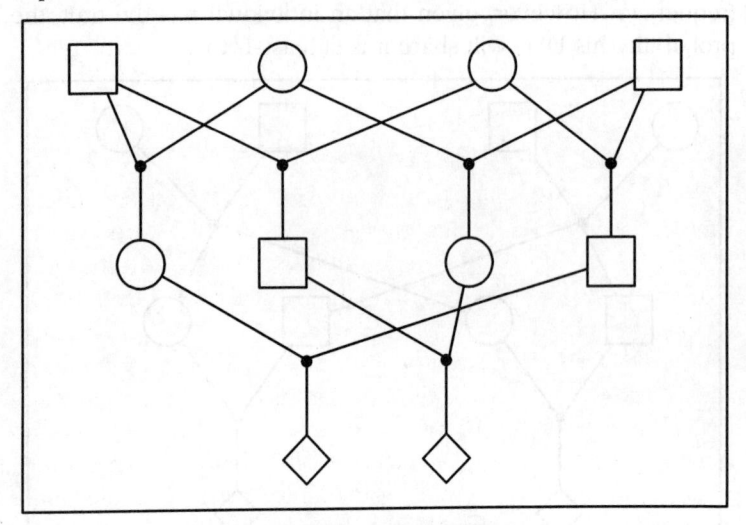

Figure 2: Quadruple Half First Cousins

Before we leave this story of simple pairwise relationships, there is one further twist. Consider a pair of (paternal) half-sibs; just like the grandparent-grandchild, at any given locus they share one gene IBD with probability 1/2. For a single genetic locus, the genetic consequences of being half-sibs is the same as that of being grandchild and grandparent (and incidentally the same as being an uncle and nephew). However, here enters genetic linkage, and the fact that genes come on chromosomes. Sharing genes IBD at a locus is clearly not independent of sharing genes IBD at a very closely linked locus; tightly linked genes tend to segregate together. Because of the different pedigree structures, and the pattern of meioses involved, patterns of genome sharing at linked loci is not the same for these three pairwise relationships[10]. Grandparent-grandchild have the biggest contiguous pieces of genome shared IBD, since only the meiosis from parent to the grandchild breaks up the haplotype transmitted from the grandparent to the parent. For half-sibs, there are two meioses to be considered. For uncle-nephew there are five, and the pattern is more complex, but there are, on average, more and smaller pieces of genome shared IBD. The larger and fewer the pieces, the larger the variance of the proportion of genome shared.

Complexity of Human Pedigrees

The simplest pedigrees on which we may analyse genetic data are relative pairs, and the simplest of these are parent-offspring pairs, sib pairs, or even twin pairs. At the other end of the scale are pedigrees like the genealogy of the entire Hutterite population of North America, 40,000 individuals, completely descended from about 80 founding individuals, living about 300 years ago. This is probably the best-documented pedigree of a human genetic isolate. One could even consider the entire genealogy of Iceland, founded 1000 years ago, or in principle, something larger yet, although accurate pedigree information would be hard to obtain. In considering entire genetic isolates or populations, we normally analyse data on pedigrees extracted from the population, not the entire population pedigree. Nonetheless, such pedigrees are often highly complex.

In terms of feasibility of analysis of genetic data on pedigrees, pedigree complexity is a more critical factor than pedigree size. There are different forms of pedigree complexity, but they all result from loops in the pedigree structure. The most immediate type of loop is the "inbreeding loop" caused by marriages of relatives. One example is shown in figure 3: here, an individual was ascertained as having a rare recessive disease and his parents being first cousins[11]. It was later recognised, partly because of the genetic data, that each of his parents was also the child of a first-cousin marriage. This raises an important practical point; in analysing genetic data, we may seek large pedigrees, but we seldom actively seek complex ones. However, in mapping disease genes for recessive traits we may seek inbred individuals, and even where we do not seek them, individuals exhibiting rare recessive traits have high probability of being inbred. Thus our pedigrees likely derive from communities, or societies, in which there is high incidence of marriage among close relatives, and pedigrees with multiple inbreeding loops will likely occur.

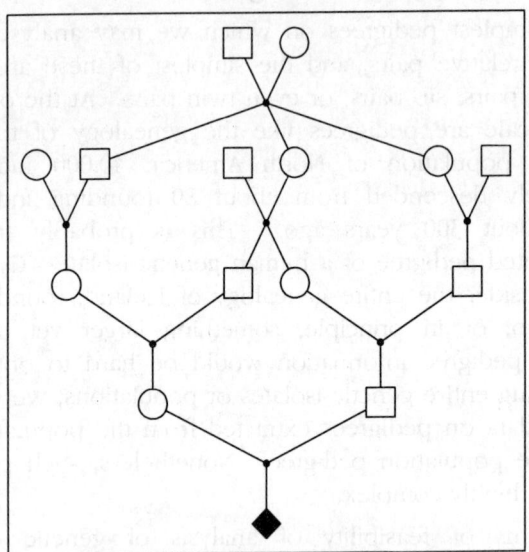

Figure 3: A Cousin Marriage Pedigree

Inbreeding loops are only apparent when pedigrees are traced back at least four generations; they are common in genetic isolates, where often pedigrees are traced back six or more generations. In human pedigrees from large populations, or traced to lesser depth, a more common type of loop is the "exchange loop". Double first cousins (figure 1) provides an example; each of a pair of siblings marries each of an unrelated pair of siblings. More generally, exchange loops arise when pairs of relatives marry pairs of relatives - clearly this will happen often in a local population. QHFC can also be considered as a more complicated example with a set of exchange loops; pairs of half-sibs marrying pairs of half-sibs. However, it is also the simplest example of the third type of loop; a "marriage ring" - A marries B marries C marries D marries A. These are, presumably, rare in human populations - I do not know of a real example. They arise routinely in animal pedigrees - for example thousands of cows may be mated to one bull this year, and another bull next year. However, some human populations, such as the Greenland Eskimos do have long chains of multiple marriages - A marries B marries C and D, and C marries E, and D marries F, who marries G.... In conjunction with other relationships among individuals or their descendants there can be many loops created by these marriage relationships[12].

Genetic Analyses of Complex Traits

In 1950s and 1960s, there were several studies of pedigrees of genetic isolates, but the scientific questions were not always clear. One example is the study of the Tristan da Cunha islanders, after their evacuation to UK in 1961, when their volcano erupted[13,14]. This was an excellent study in its accuracy and completeness; it's the one that first attracted me to genetic studies of genetic isolates. However, it was entirely opportunistic in its undertaking. Another example is the studies of the Hutterites of Saskatchewan and Alberta[15], perhaps the best example of a genetic isolate whose entire pedigree is accurately known. These earlier studies were population oriented - studying the genetic diseases of the particular population, with a view to counselling and health of that population.

In the 1970s there was a shift towards more general medical genetics, with studies in Salt Lake City of the Mormon pedigrees, where the focus was on the disease rather than on the population: for example, cancer[16] or heart disease (Williams* et al. 1979)[17]. [*: I would like digress for a brief moment to remember Dr. Roger Williams, an enthusiastic and energetic researcher into the epidemiology and genetics of cardiovascular disease, using the Mormon genealogical database, who was himself a committed member of the Church of Jesus Christ of Latter Day Saints. Sadly, Roger Williams was on the Swiss Air flight that crashed in September 1998, on his way to a meeting in Geneva.]

In 1979-80, came a sudden change, with the development of the first DNA genetic markers[18], followed, since then, by many other types of DNA markers. These opened the way for genetic linkage mapping of the entire human genome, the construction of human genetic maps, the mapping of many human disease genes, and ultimately to the Human Genome Project. Suddenly human pedigrees, particularly those in which a disease was segregating, had a major purpose, providing information for locating the gene.

Genetic epidemiology is moving towards attempting to resolve ever more complex traits by segregation and linkage analysis. There has been debate over whether one will be most successful using very small pedigrees, such as sib pairs, which may be available in large numbers, or larger pedigrees. It is generally accepted that, where available, extended pedigrees provide more power, since transmission of genes over the generations can be more easily seen, there is likely to be greater trait homogeneity, and a smaller proportion of founder individuals. A complex pedigree is not necessarily helpful, but we have seen that those extended pedigrees that are available and well studied are often complex.

One case in which complex pedigrees, at least at the level of inbred individuals, are sought, is for homozygosity mapping:

the pedigree of figure 3 was obtained in such a study. This method is directed towards mapping of genes with recessive effects, and the rationale is as follows. Individuals who are inbred and whose parents are closely related (e.g. first cousins) have high posterior probability of carrying two IBD genes at the disease locus. If they are so, then they will be IBD not just at this locus, but for a patch of genome around it. These patches will give rise to patches of homozygosity, since IBD genes must be of the same allelic type. Where the same patches are observed in different inbred affected individuals, this provides evidence for linkage.

The same rationale underlies all methods of linkage detection using affected relatives: the relatives share the trait, so have increased probability of sharing genes IBD that predispose to the trait, and hence of sharing genes IBD at closely linked marker loci, and hence of sharing alleles at these marker loci. Close relatives, such as sib pairs, share much of their genome IBD, so many sib pairs are needed to detect excess sharing at certain locations in the genome. Genetic homogeneity of the trait is then questionable. Distant relatives may be more useful; at 12-14 meioses removed from each other, there is high prior probability that these relatives will not share any of their genome IBD (Donnelly, 1983). Thus any regions that appear to be IBD in affected relatives are of interest. However, because the prior probability of IBD is small, allele sharing at a single locus carries little weight; the posterior probability of IBD will still be quite small. It is allele sharing in patches of multiple markers that builds the evidence for IBD. However, for remote relatives, patches of IBD are expected to be small. Dense marker maps are needed. There is a pay-of between the scale of the genetic map and marker density, and the degrees of relationship one should consider in trying to detect linkage from genome sharing.

Now, with the Human Genome Project nearing completion, the advent of single nucleotide polymorphisms provide the prospect of a truly dense genetic map. What will be the role of pedigree studies in using these markers to resolve human genetic traits? Currently there is much discussion about population-level associations, or "disequilibrium mapping" as an approach to mapping and identifying genes relating to human diseases. Although the exact pedigree may be unknown, the same framework of gene IBD underlies these data. The assumption is that affected individuals do share IBD alleles predisposing them to the disease or trait. It is this IBD which results in associations at the population level, just as IBD in pedigrees results in similarities among relatives. Although the pedigree relationships may be unknown to us, the gene sharing among these individuals, providing association at the population level, can provide evidence for linkage and fine-scale mapping. For these remote relatives, IBD segments will be shorter yet, and the new and yet denser genetic maps may serve a useful purpose, in the next stage of resolving human genetic traits.

References:

[1] Pearson K and Lee A (1903). *On the laws of inheritance in man. I. Inheritance of physical characters.* Biometrika 357-462.

[2] Fisher, R A (1918). *The correlation between relatives on the supposition of Mendelian inheritance.* Transactions of the Royal Society of Edinburgh, 52: 399-433.

[3] Hogben, L T (1931). *The genetic analysis of familial traits. I. Single gene substitutions.* Journal of Genetics 25, 97-112.

[4] Haldane, J B S (1934). *Methods for the detection of autosomal linkage in man.* Annals of Eugenics 6, 26-65.

[5] Fisher, R A (1934). *The amount of information supplied by records of families as a function of the linkage in the population sampled.* Annals of Eugenics 6, 66-70.

[6] Galton, F (1904). *Average number of kinfolk in each degree.* Nature 70, 529 and 626.

[7] Donnelly, K P (1983). *The probability that related individuals share some section of the genome identical by descent.* Theoretical Population Biology 23, 34-64. Fisher R A (1914). The Eugenics Review 5, 309-319.

[8] Fisher, R A (1930). *The Genetical Theory of Natural Selection.* Clarendon Press, Oxford.

[9] Thompson, E A (1990). *Inference of a gene and its paths of descent: the Newfoundland example.* American Journal of Human Biology 2, 291-301.

[10] Thompson, E A (1986). *Pedigree Analysis in Human Genetics.* The Johns Hopkins University Press, Baltimore, MD

[11] Goddard, K A B, Yu, C-E, Oshima, J, Miki, T, Nakura, J, Piussan, C, Martin, G M, Schellenberg, G D, Wijsman, E M (1996). *Towards localization of the Werner syndrome gene by linkage disequilibrium and ancestral haplotyping: lessons learned from analysis of 35 chromosome 8p11.1-21.1 markers.* American Journal of Human Genetics 58, 1286-1302.

[12] Sheehan, N A (1992). *Sampling genotypes on complex pedigrees with phenotypic constraints: the origin of the B allele among the Polar Eskimos.* IMS Journal of Mathematics Applied in Medicine and Biology 9, 1-18.

[13] Lewis, H E (1963). *The Tristan islanders a medical study of isolation.* New Scientist 20, 720-722.

[14] Roberts, D F (1971). *The demography of Tristan da Cunha.* Population Studies 25, 465-479.

[15] Hostetler, J A (1974). *Hutterite Society.* Baltimore, MD: Johns Hopkins University Press.

[16] Skolnick, M H (1977). *Prospects for population oncogenetics.* In "Genetics of Human Cancer", J J Mulvihill, R W Miller, J F Fraumeni (eds). New York, Raven Press, P.19.

[17] Williams, R R, Skolnick M H, Carmelli D, Maness A T, Hunt, S C, Hasstedt S, Reiber G. E. and Jones R K (1979). *Utah pedigree studies: design and preliminary data for premature male CHD deaths.* In "Genetic Analysis of Common Diseases: Applications to Predictive Factors in Coronary Disease", C F Sing and M H Skolnick (eds). New York, Alan R Liss Inc., Pp.711-729.

[18] Botstein, D, White, R L, Skolnick, M H, and Davis, R W (1980). *Construction of a genetic linkage map in man using restriction fragment length polymorphisms.* American Journal of Human Genetics 32, 314-331.

A Brief History Of The Pedigree In Human Genetics

Robert G Resta[1]

Background

The word "pedigree" comes from the French term *pie de grue*, meaning "the crane's foot" (OED, 1989). The word is probably derived from the observation that the curved lines used in early pedigrees to connect offspring to their parents resembled a bird's claw (Sweet, 1895). Genealogical pedigrees date back at least five centuries, arising perhaps out of a legal need to distribute inheritance to appropriate relatives.

In contrast to genealogical pedigrees, genetic pedigrees are usually drawn to demonstrate the inheritance of biological traits rather than to identify specific individuals. Although pedigrees of various types have appeared in medical journals for two centuries, the development of the pedigree as a scientific tool coincides with the rise of human genetics and eugenics during the twentieth century.

The Nineteenth Century

Throughout the nineteenth century various forms of the pedigree appeared in medical journals, usually to demonstrate that heredity was involved in the etiology of a particular illness or pathological condition such as limb deformities. However, pedigrees were not typically used to study inheritance patterns or to predict who in the family may be at risk for a particular

---

[1] An earlier version of this work appeared as Resta R G (1993) The crane's foot: The rise of the pedigree in human genetics. J Genetic Counsel 2:235-60.

disease or trait. (eg Harrington, 1885; Parker and Robinson, 1887; Ribot, 1875; Rushton, 1994), and pedigrees were not considered crucial evidence for demonstrating the heritability of a trait. For example, Huntington, in his classic paper describing Huntington disease (Huntington, 1872) did not use a pedigree. Nonetheless, he quite accurately described some genetic aspects of this neurological disorder, such as affected individuals being born only to affected parents and the lack of generation skipping. However, Huntington did not try to estimate the risk that a family member might inherit the disease. Instead, he simply stated only that "one or more of the offspring almost invariably suffer from the disease, if they live to adult age."

Figure 1: Pedigree by Pliny Earle, demonstrating inheritance of colour blindness in his own family (Earle, 1845)

In 1845, Pliny Earle, a physician, published a pedigree which used circles and squares to illustrate the inheritance of colour blindness in his own family (See Figure 1) (Earle, 1845). According to E Nettleship, a physician and a member of the Eugenics Education Society (see below for a discussion of the Eugenics Education Society), Earle's choice of style was prompted by practicality (Discussion, 1913-14). Earle was apparently unable to obtain printers' symbols other than those used for printing music; thus, unaffected females were represented by whole note symbols (minims) and affected females by blackened quarter note symbols (crotchets). However, Earle did not speculate on the inheritance pattern of the disorder.

Galton and Eugenics

Francis Galton is usually regarded as the father of eugenics. In his book *Hereditary Genius: An Inquiry Into Its Laws and Consequences*, published in 1869, pedigrees were used to demonstrate the inheritance of genius and artistic ability in famous families. This work contains familiar drop-line pedigrees in which vertical lines and horizontal lines connect generations and sibships, respectively. Galton used names of family members and a description of "genius" traits instead of symbols to indicate gender and trait status. Indeed, the purpose of Galton's pedigrees, as in genealogical pedigrees, was to identify individuals rather than to maintain anonymity. Galton was Charles Darwin's half first cousin[2] and both men were grandsons of Erasmus Darwin, the eminent English physician. In all likelihood, this family history influenced Galton's interest in eugenics.

Galton's second major book on inheritance, *Natural Inheritance* (1889), is a study of "similarities of moderately exceptional qualities in brotherhoods and multitudes" (Galton, 1889, p249). Galton solicited families who provided data (for cash) on stature, eye colour, artistic faculty, and disease. No pedigrees were presented in this study of 160 families. However, in an appendix, Galton acknowledged the problems resulting from the lack of uniformity in pedigree nomenclature and style. The pedigree technique he offered did not utilise geometric shapes (See Figure 2). Rather, individual names were entered on a page in a quarto book, divided in half for the maternal (right side) and paternal ancestry (left side). Each half page contained boxes in which the maternal and paternal ancestors were recorded, along with their initials, relationship to the proband, illnesses, cause of death and age at death.

---

[2] Galton and Darwin shared a grandfather but their parents were half-siblings

PROBAND						
Father's Father's Father and his fraternity		Father's Mother's Father and his fraternity		Mother's Father's Father and his fraternity		Mother's Mother's Father and his fraternity
Father's Father's Mother and her fraternity		Father's Mother's Mother and her fraternity		Mother's Father's Mother and her fraternity		Mother's Mother's Mother and her fraternity
Father's Father and his fraternity		Father's Mother and her fraternity		Mother's Father and his fraternity		Mother's Mother and her fraternity
spare space	Father and his fraternity	spare space	spare space	Mother and her fraternity		spare space
spare space				children		

Figure 2: Galton's Technique for Recording a Family History in a Quarto Notebook (adapted from Galton, 1889)

Pedigrees in the Twentieth Century

The growth and development of eugenics was coincidental with the rise of genetics at the turn of the twentieth century. Eugenicists tried to apply genetic principles to the solution of social problems such as poverty, prostitution and crime. Active eugenics movements developed in England and the United States. Most of the eugenic activity in England was concentrated in the Galton Laboratory of National Eugenics and the Eugenics Education Society whereas in the United States eugenic activity was centred at the Eugenics Record Office at Cold Spring Harbor.

The Galton Laboratory for National Eugenics

The Galton Laboratory for National Eugenics (initially called the Eugenics Record Office) at University College was established in 1905 by Francis Galton. Karl Pearson, pioneer statistician and student of Galton, became its director in 1907 (Farrall, 1979: Kevles, 1985; Mazumdar, 1992).

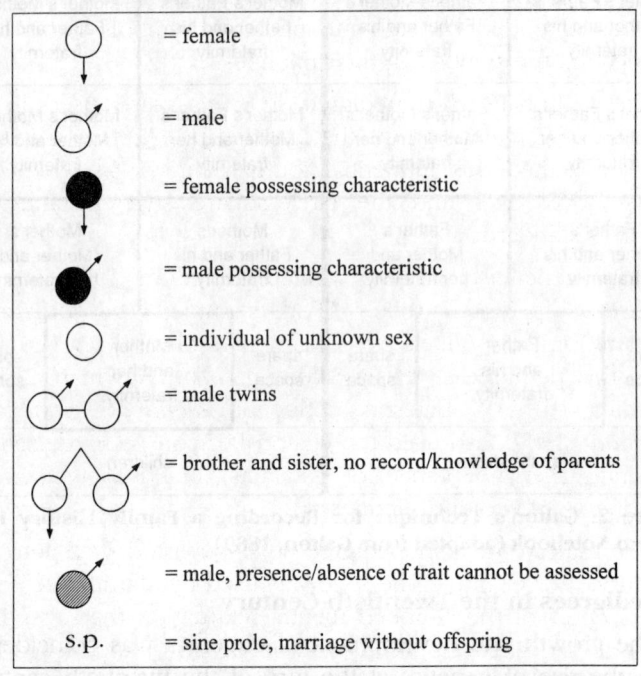

○ = female

♂ = male

● = female possessing characteristic

● = male possessing characteristic

○ = individual of unknown sex

⚤ = male twins

= brother and sister, no record/knowledge of parents

= male, presence/absence of trait cannot be assessed

s.p. = sine prole, marriage without offspring

Figure 3: Some Pedigree Symbols Used in Pearson's *Treasury of Inheritance* (adapted from Pearson, 1912)

Pearson, through the Eugenics Laboratory, published *The Treasury of Human Inheritance* (1912), a compendium of articles on the genetics of a variety of conditions, such as diabetes, polydactyly, cleft palate, legal ability, and insanity. In the introduction to this work, Pearson describes his preferred method of pedigree style and symbols, a style apparently borrowed from Galton (Davenport & Laughlin, 1915; Mazumdar, 1992). The pedigrees that appear in *The Treasury* follow Pearson's guidelines fairly closely and are very similar to the style used by most geneticists today. The major difference was the use of the Mars (♂) and Venus (♀) symbols to represent males and females, rather than squares and circles (See Figure 3). In addition, mating lines between males and females extended beneath the pedigree symbols rather than connecting the symbols at their midlines.

The Galton/Pearson pedigree style was also used in the *Annals of Eugenics* (now the *Annals of Human Genetics*). Pearson edited this influential journal until 1934 when R A Fisher took over and the circle/square style became this journal's standard.

Most geneticists in the United Kingdom followed *The Treasury*'s pedigree style. For example, *The Journal of Genetics*, initially edited by William Bateson and R C Punnett, used this pedigree style (eg Salomon, 1910; Newman, 1913-4; Hawkes, 1913-4). Bateson's influential *Mendel's Principles of Heredity* (1909) also used the Galton/Pearson pedigree style to illustrate the inheritance of cataracts, brachydactyly, night-blindness, and colour blindness.

The Galton/Pearson pedigree style persisted well into this century. For example, the style was used by Martin and Bell in their report of the X-linked mental retardation that eventually came to be called Fragile X syndrome (Martin and Bell, 1943). Arnold Sorsby, the first editor of the *Journal of Medical Genetics*, recommended that contributors use the Galton/Pearson style (see "Instructions for Contributors" in the journal's first issue in 1964). Nine years later (Volume 10, 1973), under editorship of C A Clarke, both the Galton/Pearson and the circle/square style were allowed.

The Eugenics Education Society and Pedigrees

The Eugenics Education Society (renamed the Eugenics Society in 1926) was established in London in 1907 by Sybil Gotto, the widow of a naval officer (Mazumdar, 1992). The Eugenics Education Society took an active role in popular and professional education by sponsoring eugenic lectures, setting up educational exhibits at fairs, and publishing eugenic pamphlets. Galton was elected Honorary President in 1908 (Anonymous, 1909). Pearson, however, had little patience for this society or its members (Mazumdar, 1992).

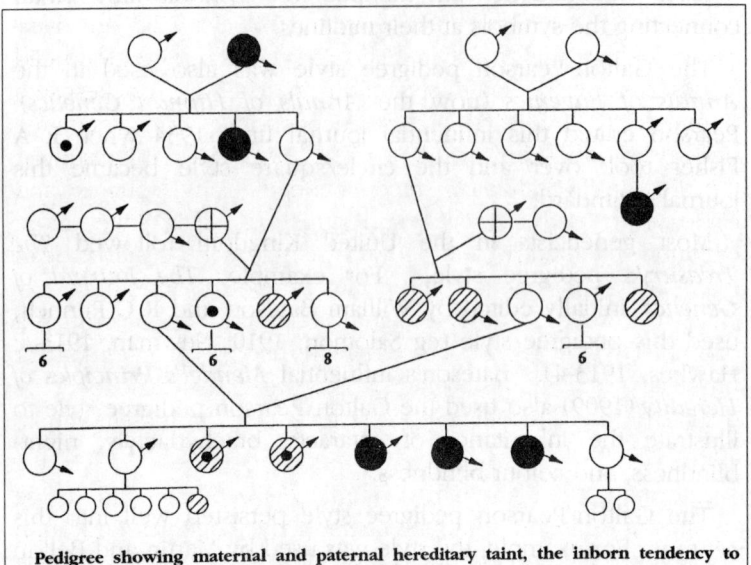

Pedigree showing maternal and paternal hereditary taint, the inborn tendency to insanity manifesting itself in the form of insanity of adolescence in their offspring.

Figure 4: Typical Eugenics Education Society pedigree, demonstrating inheritance of a dysgenic trait (adapted from Mott, 1910, Figure VIII).

The Eugenics Education Society collected many pedigrees in an attempt to demonstrate genetic aspects of criminality, pauperism, and feeble-mindedness (society's "residuum") (See Figure 4). The purpose of the pedigrees was not to determine Mendelian ratios or individual risks; rather the purpose was to show that these traits had significant genetic components (Mazumdar, 1992). Pedigree after pedigree, containing hundreds of individuals and covered with numerous blackened symbols, provided strong visual "proof" of the scope of the eugenic problem. These pedigrees were more a form of propaganda rather than a rigorous scientific tool.

The pedigree style utilised by the Eugenics Education Society was similar to the Galton/Pearson style (Lidbetter, 1910-11; Lidbetter, 1912-13), but members of the Society did not draw pedigrees in a uniform style. The Society recognised the need to standardise pedigree style and in 1912-13, its Research Committee published recommendations for pedigree standardisation (Carr-Saunders et al, 1912-13). Interestingly, the Research Committee recommended using squares and circles to represent males and females. This style, more common in America (see below), was felt to be easier to read than the style using Mars and Venus symbols (Carr-Saunders et al, 1912-13). In the early 1930s the Eugenics Society printed a pamphlet ("How To Prepare a Family Pedigree") intended for physicians and interested lay people; in it they described the details of their recommended style (Notes and Memoranda, 1931-32; Hall, 1990).

Although the Research Committee recommended using squares and circles, some members of the Eugenics Education Society objected, preferring instead the Galton/Pearson style (Discussion, 1913-14). In fact, the Galton-Pearson style remained the most commonly used pedigree style in England until the 1960s.

The Society's own *Eugenics Review* (now the *Journal of Biosocial Science*), continued to use the Galton-Pearson style through 1932. The single exception in this journal was Henry Goddard's article on the inheritance of feeble-mindedness in which he used the circle/square style (Goddard, 1911a). Goddard was a close associate of Charles Davenport, the director of the American Eugenics Record Office and the chief proponent of the square/circle style (See below).

The Pedigree in America

The Galton/Pearson pedigree style was common in America until about 1910. For example, Farabee's report on brachydactyly (Farabee, 1905), generally cited as the earliest description of an autosomal dominant trait in humans, is illustrated with a Galton/Pearson style pedigree. As in England, subsequent development of the pedigree in America and the adoption of the circle/square style were related to the rise of the eugenics movement and to the influence of individuals connected with it.

Charles Davenport and the Eugenics Record Office

Charles Davenport, an influential biologist, was the director of the prestigious Station for the Experimental Study of Evolution at Cold Spring Harbor, New York. While at Cold Spring Harbor, he established the Eugenics Record Office (ERO) and appointed Harry Laughlin as the Superintendent (Allen, 1986; Resta, 1992). The ERO became the American center for the collection and dissemination of eugenic information and ideology.

Davenport's earliest articles on human heredity contained either Galton/Pearson style pedigrees or no pedigrees at all (Davenport & Davenport, 1907; Davenport & Davenport, 1910). It was not until Davenport's interest in eugenics became his primary professional activity that pedigrees became a significant part of Davenport's work.

This interest in eugenics and pedigrees culminated in the publication *Heredity in Relation to Eugenics* in 1911. This book, the earliest definitive American eugenics text, contains hundreds of pedigrees intended to illustrate the inheritance of undesirable traits such as prostitution, pauperism, and feeble-mindedness. Amey B Eaton, Davenport's research assistant at the ERO, drew the pedigrees using squares and circles to represent males and females respectively (Davenport, 1911).

Only a few pedigrees, taken from published works of other authors, were drawn in the Galton/Pearson style.

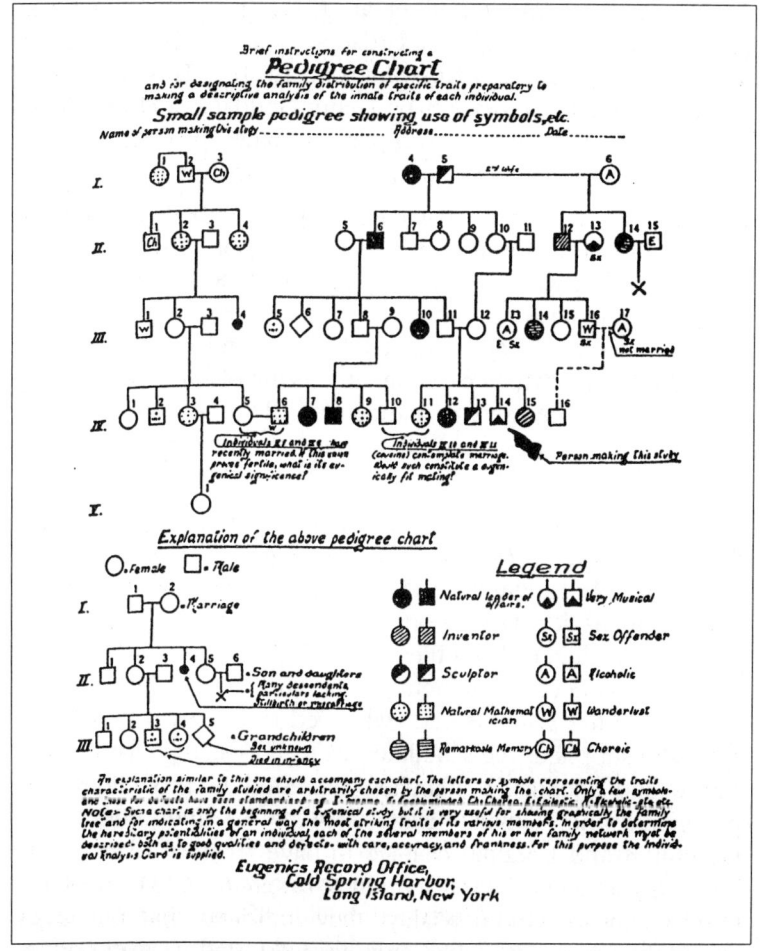

Figure 5: Typical symbols used in ERO pedigrees (from Goddard, 1911b).

Davenport claimed that the pedigree style evolved from the Galton/Pearson style (Davenport and Laughlin, 1915) but was based primarily on the recommendations of the Eugenics Committee of the American Association for the Study of the Feeble-Minded (Davenport et al, 1911) (now the American Association on Mental Retardation), although the style appeared as early as 1910 (Munson, 1910). This style was recommended to Davenport by Henry Goddard, a close friend who had worked with Davenport on other projects for the American Breeder's Association in 1910 (Gelb, 1986).

The ERO pedigree style, formally described in several publications (Davenport et al, 1911; Davenport, 1912a), is very similar to the style used by most geneticists today (See Figure 5). Besides squares and circles, ERO pedigrees contain Roman numerals to mark generations, Arabic numbers to identify individuals in each generation, and symbols for pregnancy, twins, miscarriage, and consanguinity, all of which are used today. Very large pedigrees were sometimes drawn in a circular format.

The ERO made the style available to the general public with the publication of *How To Make A Eugenical Family History* (Davenport & Laughlin, 1915). This pamphlet served as an advertisement for Davenport's eugenic research program of collecting pedigrees from the American population at large. In order to collect meaningful data, he knew he had to standardize the symbols and style used to construct pedigrees. The ERO supplied free schedules, as well as stamps for making circles and squares, to families interested in providing their pedigrees and familial characteristics for the ERO files. Davenport and Laughlin claimed to have received over 20,000 requests for schedules (Davenport & Laughlin, 1915). Sensitive to problems of confidentiality, they indicated that pedigrees would be made available only to ERO staff (Davenport & Laughlin, 1915).

Davenport believed pedigrees were important for five reasons (Davenport & Laughlin, 1915):

1) Pure love of knowledge as well as to stimulate interest in the origin of an individual's tastes, capacities, and limitations.

2) To allow society to implement effective and humane treatment of delinquent individuals.

3) To permit vocational selection based on special capacities. Davenport cites the Pierpont Morgan family as an example, claiming that only 1/100,000 individuals possessed the genetic ability to administer billions of dollars as effectively as Morgan or his son!

4) To help individuals achieve their educational potential.

5) To select proper "marriage mates."

Davenport's influence on pedigree style extended beyond the ERO. *The Journal of Heredity* (formerly *American Breeder's Magazine*), the primary outlet for eugenic articles aimed at scientists, typically utilized the ERO pedigree style (eg Blakeslee, 1914; Davenport, 1915; Osborn, 1916). Davenport was the Eugenics Secretary of the journal's parent organisation, American Genetics Association (formerly the American Breeder's Association). In *The Principles of Heredity* (Snyder, 1935) - a text dedicated to Davenport - Laurence Snyder used the ERO pedigree style to illustrate the inheritance of hemophilia, polydactyly, albinism and eye colour. Snyder's work is one of the earliest human genetics texts; he also taught the first formal course on Medical Genetics in the United States. General medical journals often used the ERO style, such as Warthin's study of genetic aspects of cancer (Warthin, 1913). Madge Macklin, one of the first American medical geneticists who also studied with Snyder, recommended the ERO style (Macklin, 1945). The two major clinical genetics journals published in the United States, *American Journal of Human Genetics* and the *American Journal of Medical Genetics*, have always used the Davenport pedigree style.

Although the ERO style was the most common pedigree style in America after 1910, Galton/Pearson style pedigrees continued to appear in some of Davenport's own work (Davenport, 1912b). In addition, the journal *Genetics*, on whose editorial board Davenport sat, utilised the Galton/Pearson pedigree style for many years (eg Cushing, 1917). General medical journals also occasionally used the Galton/Pearson style, as demonstrated by Valentine and Neel's paper which documented the autosomal recessive inheritance pattern of beta thalassemia (Valentine and Neel, 1944).

The Degenerate Family Studies

The first studies of "degenerate" families appeared in the latter half of the nineteenth century. Robert Dugdale's study of the Jukes Family (1877), the first of the so-called degenerate family studies (Rafter, 1989), contained ancestry charts but no pedigrees. Winship's follow-up study of the Jukes also did not utilize pedigrees (Winship, 1900).

During the first two decades of the twentieth century, the ERO sponsored and inspired many studies of "degenerate" families (Danielson & Davenport, 1912; Kite, 1913; Sessions, 1918; Finlayson, 1919). These studies attempted to illustrate genetic aspects of feeble-mindedness and criminality by tracing lineages with a history of "defective germ plasm" (Rafter, 1988).

It was not until Davenport became involved with these family studies in 1912 that defective lineages were traced by means of pedigrees; the style used was that of the ERO (Danielson & Davenport, 1912). Like the pedigrees collected by the Eugenics Education Society in England, the pedigrees of these "degenerate" American families made a significant impression on the public and researchers (Smith, 1985). It was also about this time that the word "pedigree" became a colloquial term for an individual's criminal record (OED, 1989).

International Perspectives

Many countries participated in the growth of eugenics and human genetics. The International Federation of Eugenics Organizations, formed in 1912, comprised of representatives of eugenics organisations from over 20 countries (Laughlin, 1934; International Federation of Eugenics Organizations, 1934). Its Committee on Standardization of Pedigree Charts recommended adoption of the ERO pedigree style, with minor modifications (Notes and Memoranda, 1926). This recommendation is not surprising, since one of the two committee members was Harry Laughlin, the Superintendent of the ERO.

The International Federation was responsible for organising the three International Congresses of Eugenics held in London (1912) and New York (1921,1932). These congresses served as scientific forums for the presentation and discussion of eugenical studies. Numerous pedigrees were displayed at these eugenic congresses and, in general, the styles diverged along national lines: Americans followed the ERO pedigree style and Europeans used the Galton/Pearson pedigree style. (*Eugenics Education Society*, 1912; Van Wagener, 1912; Weeks, 1912; Mott, 1912; *Eugenics, Genetics and the Family*, 1923; Banker, 1923; Mjoen, 1923; Ruzicka, 1923; *A Decade of Progress*, 1934).

In addition to the work of the International Federation, Germany supported a program in genetics and eugenics in the opening decades of the twentieth century (Weiss, 1987). Like America and England, pedigrees were an important feature of German eugenic studies of the *Minderwertigen* ("the less valuable"). German eugenicists and geneticists employed both the Galton/Pearson style (Baur, 1922; Baur, Fischer, & Lenz, 1931) or the ERO style (Lidbetter et al, 1912-13).

German geneticists utilised some alternative techniques when depicting family ancestry. The Ahnentafel was a horizontal depiction of a lineage in which no symbols were used, but names and descriptions were written directly on the chart (Mazumdar, 1992). The *Sippschaftstafel* was a dramatic pedigree form in which the proband was placed in the center, and maternal and paternal lineages radiated in curved lines from the proband. This type of pedigree was supposed to demonstrate the proband being crushed by the weight of his or her dysgenic ancestry (Mazumdar, 1992). The *Sippschaftstafel* presented by Ernst Rüdin in 1911 used squares and circles to depict gender.

Discussion

Geneticists in the early twentieth century adapted legal and genealogical pedigree styles for purposes of scientific investigation and to demonstrate eugenic arguments. The pedigree techniques developed by these geneticists reflected national styles (circle/squares in America, Mars/Venus symbols in England). The national pedigree styles to some extent followed the theoretical constructs of genetics that were most popular in those countries. The pedigree style adopted in each country reflected the personal preferences of important figures in the genetics and eugenics communities - Galton, Pearson and Davenport. These scientists established standards of pedigree construction which are still more or less followed today.

The superficial differences in pedigree styles were not as important as the similarity of the information contained in ERO and Galton/Pearson pedigrees. The information content of pedigrees, such as what is or is not a biological trait and which traits are worthy of scientific investigation, reflected the research agenda of geneticists. Eugenics was the primary research interest of those geneticists who established the pedigree as an analytic tool. Thus, the information content of pedigrees often reflected the interests and biases of eugenicists, who claimed that many good and bad behavioural traits had a significant genetic component.

The pedigree served both explicit and implicit functions for eugenicists. Explicitly, a pedigree was purported by eugenicists to be an objective tool for conveying genetic information about dysgenic families. Carefully blackened circles and squares were, in eugenicists eyes, scientific data points. In turn, this objective display of data helped legitimise human genetics and eugenics as "real science."

Implicitly, pedigrees permitted eugenicists to objectify the families they studied. Reducing their subjects to geometric shapes reinforced the notion that the dysgenic families with their problems of feeble-mindedness, poverty, and squalid living conditions were somehow less than human. Once rendered non-human, dysgenic families were no longer worthy of social and economic support. Betraying eugenics' historical roots in the American Breeders Association, some eugenicists treated dysgenic families as if they were animals whose mating required scientific intervention and regulation. In addition, the powerful visual impact of pedigrees covered with "defective" traits helped persuade the general public of the scope of the eugenic problem.

Many scientists involved with eugenics were also leading figures in the rise of human genetics in the early twentieth century. Until the 1930s, when mainstream eugenics fell into disfavour, the distinction between eugenics and human genetics was unclear. The same scientists published in both human genetics and eugenics journals and presented papers at both eugenics and human genetics conferences. Membership in both human genetics and eugenic organisations overlapped, and in the case of the Italian Eugenic and Genetic Society, coincided (Gini, 1934). Papers about polydactyly and cleft lip appeared in journals and at conferences side by side with papers about pauperism and feeble-mindedness. And the same tool, the pedigree, was used by eugenicists and human geneticists to demonstrate or prove their hereditarian claims.

The pedigree is still a vital tool for geneticists. The Human Genome Project, a federally funded endeavour to map the entire human genome, relies heavily on pedigrees. Genetic and medical journals continue to publish numerous pedigrees. New genetic technology such as pre-symptomatic DNA testing and assisted reproduction demand new nomenclature, and the need for further enhancement and standardisation of pedigree symbols (Bennett et al, 1993; Bennett et al, 1995).

Pedigrees continue to reflect social and ethical implications of genetic knowledge, along with concepts of family and genetic disease (Nukaga and Cambrosio, 1997). One of the primary concerns with the publications of pedigrees is maintaining patient confidentiality while still trying to construct a pedigree that contains scientifically meaningful information (Powers, 1993; Botkin et al, 1998). In an attempt "disguise" pedigrees, some authors have taken to "neutering" pedigrees by no longer distinguishing between males and females such that all individuals are depicted with unisex diamonds, rather than squares and circles (Levy-Lahad et al, 1995; Adam et al, 1998).

This anonymity is in sharp contrast to, and perhaps a reaction to, pedigrees in which family names or initials were included in published pedigree (such as the pedigrees by Earle, Parker and Robinson, and Valentine and Neel mentioned above). In perhaps the most extreme case of breach of confidentiality, the disease itself is named after a family, such as Christmas disease (hemophilia B). The authors of the first report of described Hemophilia B felt that long-standing medical tradition justified naming the disease after the first patient described with the disorder:

The naming of clinical disorders after patients was first introduced by Sir Jonathan Hutchinson and is now familiar from serological research....

Biggs et al., 1952, p.1379

These authors felt that this technique for naming diseases helped avoid the ptifalls of the then popular approach of naming diseases after the presumed etiology. However, the presumed etiology of some diseases was sometimes later found to be mistaken, and thus the name of the disease would reflect inaccurate information, and would become inappropriate. Imagine the uproar that would have occurred in the 1980s if AIDS had been named after the first victim of the disease!

While patient confidentiality is certainly extremely important, geneticists may pay the price of maintaining confidentiality with the currency of lost clinical information. Pedigrees that omit sex, age and other potentially relevant familial information could impair the ability of geneticists to accurately interpret new discoveries and re-interpret previous information. For example, individuals with Huntington disease who inherit a mutation from their father have a different average age of onset of symptoms than individuals who inherit the mutation from their mother (Snell et al., 1993). Such a discovery could have been lost or delayed if published Huntington disease pedigrees were "neutered" and "ageless."

Geneticists of today are not morally superior to their predecessors, nor does it seem likely that twenty-first century geneticists, with their vast array of new discoveries, will hold an ethical upper hand over their forebears. No doubt pedigrees will continue to obliquely mirror the ethical struggles and personal biases of the very human geneticists who construct these symbolic representations of genetic relationships.

Bibliography:

A Decade of Progress in Eugenics: Scientific Papers of the Third International Congress of Eugenics, New York, 1932. (1934) Baltimore: Williams & Wilkins.

Adams L J, Mitchell P B, Fielder S L, Rosso A, Donald J A, Schofield P R (1998) *Amer J Human Genet* 62:1085-1091.

Allen G E (1986) The Eugenics Record Office at Cold Spring Harbor, 1910-1940: An essay in institutional history. *Osiris* (2nd series), 2:225-264.

Anonymous (1909) Origins and work of the Society. *Eugenics Rev* 1:50-54

Banker H J (1923) The learned blacksmith. in *Eugenics, Genetics and the Family*. Baltimore, Williams and Wilkins, pp.340-347.

Bateson W (1909) *Mendel's Principles of Heredity*. Cambridge: University Press.

Baur E (1922) *Vererbungslehre*. Berlin: Verlag von Gebruder Borntraeger.

Baur E, Fischer E, Lenz F (1931) *Human Heredity*. Translated by Eden and Cedar Paul, New York: Macmillan and Co.

Bennett R L, Steinhaus K A, Uhrich S B, O'Sullivan C K (1993) The need for developing standardized family pedigree nomenclature. *J Genetic Counsel* 2:267-279.

Bennett R L, Steinhaus K A, Uhrich SB, O'Sullivan C K, Resta R G, Lochner-Doyle D, Markel D S, Vincent V, Hamanishi J (1995) Recommendations for standardized human pedigree nomenclature. *Amer J Hum Genet* 2:235-60.

Biggs R, Douglas A S, Macfarlane R G, Dacie J N, Pitney W R, Merksey C, O'Brien J R (1952) Christmas disease: A condition previously mistaken for hemophilia. *Brit Med J* 2:1378-1382.

Blakeslee A F (1914) Corn and men. *Jour Heredity* 5:511-518.

Botkin J R, McMahon W M, Smith K R, Nash J E (1998) Privacy and confidentiality in the publication of pedigrees: A survey of investigators and biomedical journals. *Jour Amer Med Assoc* 279:1808-12.

Carr-Saunders A M, Greenwood M, Lidbetter E J, Schuster E H J, Tredgold A F (1912-13) The standardization of pedigrees: A recommendation. *Eugenics Rev* 4:383-390.

Castle W E, Coulter J M, Davenport C B, East E M, Tower W L (1912) *Heredity and Eugenics*. Chicago: Univ Chicago Press.

Cushing H (1917) Hereditary anchylosis of the proximal phalangeal joints (symphalangism). *Genetics* 1:90-106.

Danielson F H, Davenport C B (1912) *The Hill Folks: Report on a Rural Community of Hereditary Defectives*. Eugenics Record Office, Bulletin 1, Washington, D.C.: Carnegie Inst.

Davenport C B (1911) *Heredity in Relation to Eugenics*. New York: Holt & Co.

Davenport C B (ed) (1912a) *The Family-History Book*. Eugenics Record Office, Bulletin 7, Washington, D.C.: Carnegie Institute.

Davenport C B (1912b) The inheritance of physical and mental traits of man and their application to eugenics. in Castle W E, Coulter J M, Davenport C B, East E M, Tower W L (1912) *Heredity and Eugenics*. Chicago: Univ Chicago Press.

Davenport C B (1915) A dent in the forehead. *Jour Heredity* 6:163-164.

Davenport C B, Laughlin H H (1915) *How To Make A Eugenical Family History*. Eugenics Record Office, Bulletin 13, Washington, D.C.: Carnegie Inst of Washington.

Davenport C B, Laughlin H H, Weeks D F, Johnstone E R, Goddard H H (1911) *The Study of Human Heredity*. Eugenics Record Office, Bulletin 2, Washington, D.C.: Carnegie Inst of Washington.

Discussion - Standardization of pedigrees. (1913-14) *Eugenics Rev* 5:66-67.

Dugdale R (1877) *The Jukes*. G P Putnam and Sons, NY and London, 4th edition, Reprint edition, Arno Press, Inc (1970).

Earle, P (1845) On the inability to distinguish colours. *Am J Med Sci* IX:346-354.

Eugenics and the Family. Scientific Papers Presented at the Second International Eugenics Congress, New York, 1921. (1923) Baltimore: Williams.

Farrarall L A (1979) The history of eugenics: a bibliographical review. *Annals of Science* 36:111-

Farabee W C (1905) Inheritance of digital malformations in man. *Papers of the Peabody Mus of Amer Archaeology and Ethnology*, Harvard University. 3:69-77.

Finlayson A W (1916) *The Dack Family: A Study in Hereditary Lack of Emotional Control*. Eugenics Record Office, Bulletin 15, Washington D.C.: Carnegie Inst.

Galton F (1869) *Hereditary Genius: An Inquiry Into Its Laws and Consequences*. New York: Horton Press (1952 edition).

Galton F (1889) *Natural Inheritance*. London: MacMillan & Co.

Gelb S A (1986) Henry H Goddard and the immigrants, 1910-17: the studies and their social context. *Hist Behav Sci* 22:324-332.

Gini, Corrado (1934) Response to the presidential address. in *A Decade of Progress in Eugenics. Scientific Papers of the Third International Congress of Eugenics*. Baltimore: Williams and Wilkins, pp 23-28.

Goddard H (1911a) Heredity of feeble-mindedness. *Eugenics Rev* 3:46-60.

Hall L A (1990) Illustrations from the Wellcome Institute Library: The Eugenics Society archives in the Contemporary Medical Archives Center. *Medical History* 34:327-333.

Harrington H L (1885) A family record showing the heredity of disease. *The Physician and Surgeon* 7:49-51.

Hawkes O A M (1913-4) On the relative lengths of the first and second toes of the human foot, from the point of view of occurrence, anatomy, and heredity. *Jour of Genetics* 3:248-274.

Huntington G (1872) On chorea. *The Medical and Surgical Reporter* 26:317-321.

International Federation of Eugenics Organizations (1934) Appendix II in *A Decade of Progress in Eugenics. Scientific Papers of the Third International Congress of Eugenics.* Baltimore: Williams and Wilkins, pp 522-526.

Kevles D J (1985) *In the Name of Eugenics: Genetics and the Uses of Heredity.* New York: Alfred A Knopf.

Kite E S (1913) The 'Pineys.' published in Rafter N H (1988), *White Trash*, Boston: NE Univ Press, pp 164-184.

Laughlin H H (1934) Historical background of the Third International Congress of Eugenics. in *A Decade of Progress in Eugenics.* Baltimore: Williams & Wilkins, pp 1-14.

Levy-Lehad E, Wasco W, Poorkah P, Romano DM, Oshima, J, Pettingell W H, et al. (1995) Candidate gene for the chromosome 1 familial Alzheimer's disease locus. *Science* 269:973-977.

Lidbetter E J (1912-1913) Nature and nurture - A study in conditions. *Eugenics Rev* 4:54- 73.

Macklin M T (1945) New symbols for pedigree charts. *J Hered* 36:222-224.

Martin J P, Bell J (1943) A pedigree of mental defect showing sex-linkage. *J Neurol Pscyhiatry* 6:151-157.

Mazumdar P M H (1992) *Eugenics, Human Genetics, and Human Failings.* London, New York: Routledge.

Mjoen J A (1923) Harmonic and disharmonic race crossings. in *Eugenics, Genetics and the Family.* Baltimore: Williams and Wilkins, pp 41-61.

Mott F W (1912) Heredity and eugenics in relation to insanity. in Eugenics Education Society, *Problems in Eugenics.* London: Knight, pp 400-428.

Munson J F (1910) An heredity chart. *New York Medical Journal* 91:437-438.

Newman H H (1913-4) Five generations of congenital stationary night-blindness in an American family. *Jour of Genetics* 3:25-37.

Notes and memoranda (1926) *Eugenics Rev* 17:247-250.

Notes and memoranda (1931-32) *Eugenics Rev* 23:155.

Nukaga Y, Cambrosio A (1997) Medical pedigrees and the visual production of family disease in Canadian and Japanese genetic counselling practice. in Elson M A (ed) (1997) *The Sociology of Medical Science and Technology*. Walden, Massachusetts: Blackwell Publishers.

Oxford English Dictionary (2nd ed), (1989), Volume XI, Oxford: Clarendon Press.

Osborn D (1916) Inheritance of baldness. *Jour Heredity* 7:347-354.

Parker R W, Robinson H B (1887) A case of inherited congenital malformation of the hands and feet: Plastic operation on the feet: with a family tree. *Trans Clin Soc London* 20:181-189.

Pearson K (ed) (1912) *The Treasury of Human Inheritance*, Pts. I & II. London: Dulau and Co.

Powers M (1993) Publication-related risks to privacy: Ethical implications of pedigree studies. *Institut Review Board* 15:7-11.

Problems in Eugenics: Papers Presented to the First International Eugenics Congress, London, 1912. (1912) London: Knight.

Rafter N H (1988) *White Trash The Eugenic Family Studies, 1877-1919*. Boston: Northeastern Univ Press.

Resta R G (1992) The twisted helix: an essay on eugenics, genetic counselors, and social responsibility. *J Gen Couns* 4:227-243.

Ribot T (1875) *L'heredite Psychologique*. (10th ed - 1914), Paris, Libraire Felix Alcan.

Ruzicka V (1923) The significance of causal research in eugenics. in *Eugenics, Genetics, and the Family*. Baltimore: Williams and Wilkins, pp 449-451.

Rushton A (1994) *Genetics and medicine in the United States, 1800-1920*. Batlimore:The Johns Hopkins University Press.

Salaman R N (1910) Heredity and the Jew. *Jour of Genetics* 1:273-292.

Sessions M A (1918) The feeble-minded in a rural county of Ohio. Bulletin #6, Publication #12, Bureau of Juvenile Research, Ohio Board of Administration.

Smith D J (1985) *Minds Made Feeble: The Myth and Legacy of the Kallikaks*. Rockville, MD: Aspen Systems Corporation.

Snell R G, MacMillan J C, Cheadle J P, Fenton I, Lazrus P L, Davies P et al. (1993) Relationship between trinucleotide repeat expansion and phenotypic variation in Huntington disease. *Nature Genetics* 4:393-397.

Snyder L E (1935) *The Principles of Heredity*. D C Heath & Co: USA.

Sweet C (1895) Pedigree. *Athenaeum* March 30:409.

Valentine W N, Neel J V (1944) Hematologic and genetic study of the transmission of thalassemia (Cooley's anemia; Mediterranean anemia). *Arch Int Med* 74:185-196.

Van Wagener B (1912) Preliminary report of the Committee of the Eugenics Section of the American Breeders Association to study and to report on the best practical means for cutting off the defective germ plasm in the human population. in Eugenics Education Society, *Problems in Eugenics*. London: Knight, pp 460-479.

Warthin A S (1913) Heredity with reference to carcinoma. *Arch Intern Med* 12:546-555.

Weeks D F (1912) The inheritance of epilepsy. in Eugenics Education Society, *Problems in Eugenics*. London: Knight, pp 62-99.

Weiss S F (1987) *Race, Hygiene, and National Efficiency*. Berkeley: Univ. of California Press.

Winship, A E (1900) *Jukes-Edwards A Study in Education and Heredity*. R L Myers and Co.: Harrisburg, PA.

Computers for Research, Storage and Presentation of Family Histories

David Hawgood

There are millions of family historians using computers to store pedigrees, print family trees, find information, and organise data they have obtained. My objective is to tell you how these family historians use computers.

Genealogy software - storage of information

There are many genealogy computer packages available commercially which are good at storing and presenting family histories, and aiding research. The choice is sufficient to make it unnecessary for the genealogist to write programs, or construct a relational database. The genealogist enters information about the people and family links, also the references to the sources of information. The computer will then produce family trees and charts, with a complete set of footnotes about sources. Descriptions and reviews of genealogy packages, with lists of suppliers, are given in the books and journals listed in Reference 1.

Pedigree information is lineage-linked. A computer system to handle it has to store information about the individual, and information to tell us which other people are related. To print a pedigree from the information, the computer has to follow these links over many generations. Although computers prefer neat and tidy families, genealogy packages have to recognise the real world. Sometimes a family history includes adoptions, foster parents, multiple marriages, illegitimate children, inconsistent and approximate information. The software available has evolved over the years to handle more of the situations found in real families.

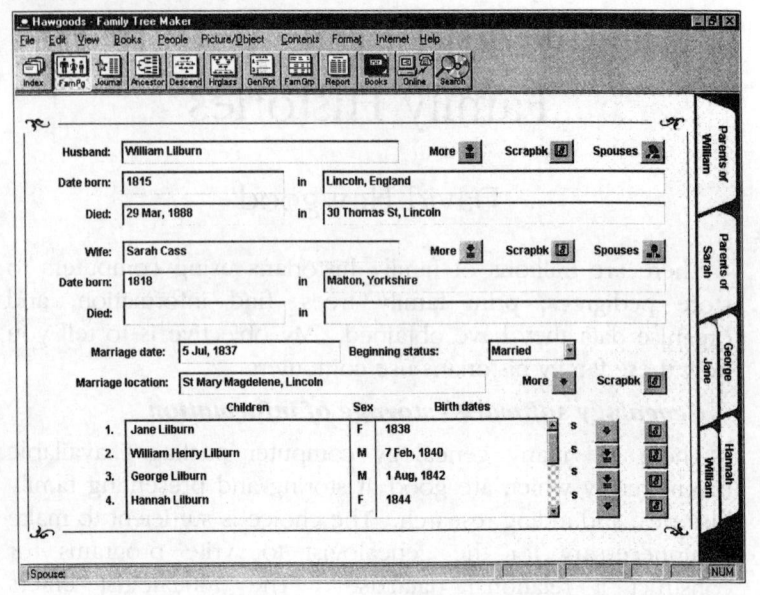

Figure 1. Family data entry screen of Family Tree Maker

Figure 1 is a data entry screen from the package Family Tree Maker, published in the USA by Broderbund. This basic family screen is displayed when the software loads, and the user just types in information about a couple - names, then date and place of birth, marriage, and death. Names and birth dates of their children can be added on this screen. The information here is about my Lilburn ancestors in Lincolnshire.

From the family page for one couple, the user can move to pages for their parents, or their children; to do this, use the mouse to click on the tabs at the side of the screen. This way the user can browse up and down the pedigree, and add extra people.

The buttons marked "More" lead to subsidiary screens. One is the Lineage page. Here the user can enter the information

USE OF COMPUTERS FOR GENEALOGICAL RESEARCH 87

about the type of relationship to parents - to cater for adoption, fostering, unknown parents. The user can also specify whether the person is to be included on family trees, and whether the person is to be included in calendars of family birthdays and anniversaries. An example where I entered information using this screen was my grand-father's illegitimate half-brother James Almond; his birth certificate leaves the father's name blank. Incidentally, this James Almond appears in the 1851 census for Blackburn Lancashire, age 8, occupation cotton power loom weaver - finding details of family history makes social history come to life. Family Tree Maker also has a page of extra facts, and a page of medical information - height, weight, cause of death, other medical information. There is also a page for entry of textual family history.

The button marked "Srapbk" leads to a scrapbook page, for linking of scanned family photographs, sound recordings, and video clips. Storing photographs in the computer and printing them as part of a family tree has become very popular with family historians.

The button marked "Spouses" leads to pages for addition of several spouses or partners, with details of the marriage, and sometimes divorce.

In some packages a particular type of fact can occur several times, each with its own source. This may be because there are several conflicting sources, eg for birth date. It may also be because there are several occupations to enter. To show this, and the way sources are entered, I am using another package, Family Origins, also published by Broderbund. The table (Fig 2) shows extra facts with three occupations for William Lilburn, as a part time policeman, as a cordwainer, and at the end of his career as a police inspector.

The columns after the name/place column show if notes and sources have been entered for the fact. For example, the information about William Lilburn being a part-time policeman was given in a newspaper article when he retired 42 years later. I entered the source reference as the Lincolnshire Chronicle, citation within that the issue for 19th October 1877, page 5 column 4, and as repository where I consulted it Lincoln Public Library.

Fact	Date	Name/Place	Note	Source
Birth	1815/16	Lincoln, England		
Chr	2 Mar 1816	St Swithin, Lincoln		X
Occupation	1835	Police constable (supernumary)	X	X
Marriage	5 Jul 1837	Sarah Cass	X	X
Occupation	1841	cordwainer		X
Occupation	1877	police detective inspector	X	X
Death	29 Mar 1888	30 Thomas St, Lincoln	X	X
Burial	31 Mar 1888	Canwick Rd Burial Ground, Lincoln	X	X

Figure 2. Facts for the life of William Lilburn, as entered in Family Origins

So far I have shown information which is all about one linked family. It is also possible to enter information about other families, or individuals, in the same database. The genealogy packages all have ways of selecting by a set of criteria - everyone named James Hawgood born after 1840, for example; sorting the resultant set; and printing a list. Thus, the package can be used for research, organising information as it arrives, as well as storing the proven linked pedigree. Most genealogists keep two databases, one for the proven pedigree and one for the miscellaneous records. The latter may extend to a complete One Name Study, collecting all occurrences of a surname, first in a locality, extending to a global study. I keep all this type of information in genealogy packages, some other genealogists prefer to keep unlinked information in a general purpose database like Access, or even in a spreadsheet.

Once the information has been entered it can be displayed or printed in a number of different formats, including some suitable for display on the World Wide Web. I will return to these later.

Transfer of data between computer systems

After entering all this information, the user may want to change to a different genealogy package, with better research facilities or a new style of chart. Or a newfound cousin from Canada has extra information held in a different genealogy package. How can the complicated lineage-linked information be moved from one computer system to another? The answer is GEDCOM, standing for GENealogical Data COMmunication. This is a standard file format, originated by the Church of Jesus Christ of Latter Day Saints (the LDS Church, the Mormons). GEDCOM is now built into almost every genealogy package. "Export" from one package produces a simple text file with tags to show the types of information - Fig 3 is an example.

Extract from a GEDCOM file	
1 BURI	
2 DATE	31 Mar 1888
2 PLAC	Canwick Road Burial Ground, Lincoln
2 NOTE	Grave space 591
1 OCCU	cordwainer
2 PLAC	West Bight, Lincoln
1 OCCU	police detective inspector
2 PLAC	Lincoln
2 NOTE	Lincoln City Watch Committee 15 Oct 1877, minutes at Lincs Archives

Figure 3. Part of a GEDCOM file with information about William Lilburn

BURI is the tag for burial, OCCU for occupation, for example. There is a complicated set of pointers in the text to link parents to children. This file can be copied, transmitted over the Internet, even modified by a word processor. Then it can be imported into a different genealogy package. All the individuals, their family linkages, and the information sources appear in the new system. Because different genealogy packages cater for different types of extra information, not all the data transfers in a simple way. For example, I said that Family Tree Maker caters for medical information, but most genealogy packages do not, so this information would often end up in a general Notes field, or in an exception listing. But the basic information of names, and date and place of vital events, usually transfers smoothly. For description of the GEDCOM format, see my book "GEDCOM Data Transfer" (ref 2) or the standard available from the LDS Church (ref 3).

Because GEDCOM is a simple and widely available standard, it is also being used for collection, storage, and distribution of pedigree data. For example, I sent my Hawgood pedigree to the LDS Church Family History Library in Salt Lake City as a GEDCOM file. It is stored in their Ancestral File of pedigrees, and distributed on CDROM to libraries world-wide. There are various libraries of pedigrees held on the Internet, with submission and downloading via GEDCOM files. There is more about libraries of pedigrees below.

Presentation of family histories by computer

Once family information has been entered, genealogy packages have a great variety of ways of displaying or printing it. The styles go from text alone, through text with linking lines, to graphics with fancy boxes and even pictures of leafy trees. The content may be full information about one person, summary lists of names and dates for many people, a family group sheet showing details for a couple and their children, a chart of descendants of one person, or a birth brief showing the pedigree of ancestors of one person. As well as variants on these, there are charts showing how two people are related, timeline charts showing the lives of people as lines against a scale of years, and others.

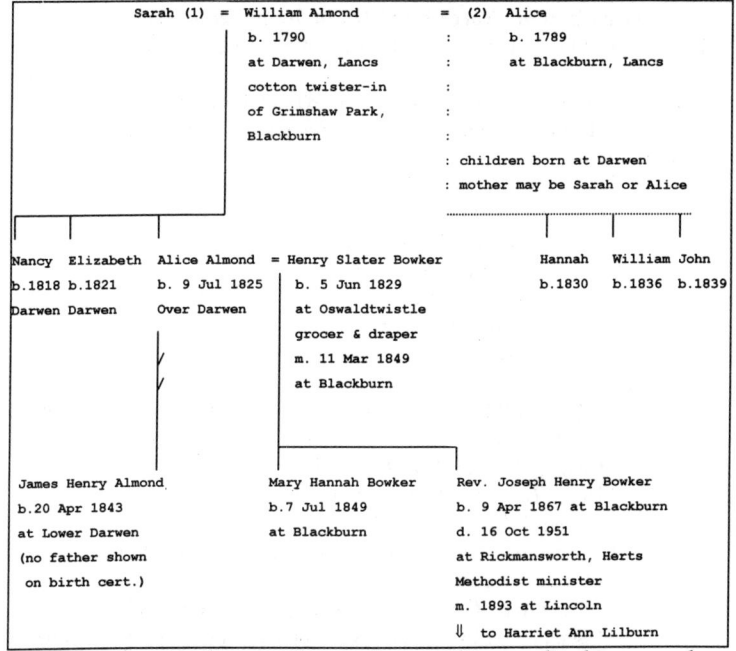

```
                    Sarah (1)  =  William Almond       =  (2)  Alice
                                  b. 1790              :       b. 1789
                                  at Darwen, Lancs     :       at Blackburn, Lancs
                                  cotton twister-in    :
                                  of Grimshaw Park,    :
                                  Blackburn            :
                                                       :
                                                       : children born at Darwen
                                                       : mother may be Sarah or Alice

 Nancy     Elizabeth   Alice Almond  = Henry Slater Bowker     Hannah    William   John
 b.1818    b.1821      b. 9 Jul 1825   b. 5 Jun 1829           b.1830    b.1836    b.1839
 Darwen    Darwen      Over Darwen     at Oswaldtwistle
                                       grocer & draper
                                       m. 11 Mar 1849
                                       at Blackburn


 James Henry Almond     Mary Hannah Bowker    Rev. Joseph Henry Bowker
 b.20 Apr 1843          b.7 Jul 1849          b. 9 Apr 1867 at Blackburn
 at Lower Darwen        at Blackburn          d. 16 Oct 1951
 (no father shown                             at Rickmansworth, Herts
 on birth cert.)                               Methodist minister
                                              m. 1893 at Lincoln
                                              ⇓ to Harriet Ann Lilburn
```

Figure 4. Drop-line family tree from Pedigree, edited in a word processor

Fig 4 shows a drop line family tree; it shows Alice Almond's illegitimate son James, then her subsequent marriage to Henry Slater Bowker. It also shows Alice's father William Almond, with two wives, but some uncertainty about which children are from which wife. Initially I produced a tree with all the information needed in a package called Pedigree published by Pedigree Software in England. Then I edited the tree in a word processor to show illegitimacy and uncertainty. An alternative to using a word processor to edit the output from a genealogy package is to use the package TreeDraw, a graphics editor for genealogy published by SpanSoft in Scotland.

> ***Descendants of George TAYLOR of Spridlington, Lincolnshire***
> 1 George TAYLOR, turnpike road labourer of Spridlington, Lincs b. 1811 d. 16 Aug 1871
> + Susannah ROBINSON b.1805 m. 12 Dec 1831 d. 29 Mar 1874
>   2 Robert TAYLOR farmer of Cammeringham, Lincs b. 1831
>   + Mary b. 1839
>   3 George TAYLOR b. 1867
>   3 Harriett TAYLOR b. 1870
>   3 John William TAYLOR b. 1875
>   3 Mary Ellen TAYLOR b. 1878
>   3 Rose Annie TAYLOR b. c.Sep 1880
>   2 Mary Ann TAYLOR b. Oct 1838
>   2 Harriet Ann TAYLOR pastry shop keeper of Lincoln b. 11 Apr 1841 d. 20 Mar 1928
>   + William Henry LILBURN sub post-master of Lincoln b. 07 Feb 1840 m. 05 Aug 1867 d. 25 Feb 1929
>   3 George William LILBURN b. 1869 d. Sep 1871
>   3 Henry Taylor LILBURN d. Oct 1871
>   3 Susanna LILBURN b. 1872 d. 1944
>   + William Bennett ROBINSON press artist, Illustrated London News b. 05 Feb 1870 m. 1896 d. 1925
>   3 Harriet Ann LILBURN b. 03 Aug 1874 d. 18 Aug 1939
>   + Joseph Henry BOWKER Methodist minister of Lincoln, Bradford, etc b. 09 Apr 1867 m. 27 Aug 1893 d. 16 Oct 1951
>   3 Alice LILBURN b. c.1879 d. 1944
>   + Frederick MYERS m. c.1904
>   2 George TAYLOR wheelwright of North Wingfield, Derbyshire b. 1847
>   + Mary SMITH b. 1846/7
>   3 Sarah B TAYLOR b. 1869/70
>   3 Harriett A TAYLOR b. 28 Sep 1880
>
> ***Figure 5. Indented descendants chart, from George Taylor of Spridlington***

Fig 5 is an indented descendants chart, very compact as a quick reference in the record office, but not very helpful to show to members of the family. Most packages produce this type of chart. Fig 6 is an ancestors chart, produced automatically by Family Tree Maker after I chose which pieces of information to print, the title and footnote, the style of box, frame and line. For variety, Fig 7 is a fan-style ancestors tree, this one was produced by PAFMate (Ref 4). Fan-style trees, if

USE OF COMPUTERS FOR GENEALOGICAL RESEARCH 93

well done, can compress a great deal of information into the page while keeping the relationships clear.

Ancestors of Harriet Anne Lilburn of Lincoln, England

```
                                                        William Lilburn
                                                        b: 1815 in Lincoln, England
                                                        m: 5 Jul, 1837 in St Mary Magdelene, Lincoln
                                                        d: 29 Mar, 1888 in 30 St Thomas St, Lincoln

                          William Henry Lilburn
                          b: 7 Feb, 1840 in Lincoln, England
                          m: 5 Aug, 1867 in St Nicholas, Lincoln
                          d: 25 Feb, 1929 in 3 Bailgate, Lincoln

                                                        Sarah Cass
                                                        b: 1818 in Malton, Yorkshire

Harriet Ann Lilburn
b: 3 Aug, 1874 in 93 Bailgate, Lincoln
m: 27 Aug, 1893 in Methodist Chapel, Lincoln
d: 18 Aug, 1939 in Rickmansworth, Hertfordshire
                                                        George Taylor
                                                        b: 1811 in New York, nr Horncastle, Lincolnshire
                                                        m: 12 Dec, 1831 in Ingham, Lincolnshire
                                                        d: 16 Aug, 1871 in Spridlington, Lincolnshire

                          Harriet Ann Taylor
                          b: 11 Apr, 1841 in Spridlington, Lincolnshire
                          d: 20 Mar, 1928 in Lincoln

                                                        Susannah Robinson
                                                        b: 1805 in Ludford, Lincolnshire
                                                        d: 29 Mar, 1874 in Lincoln
```

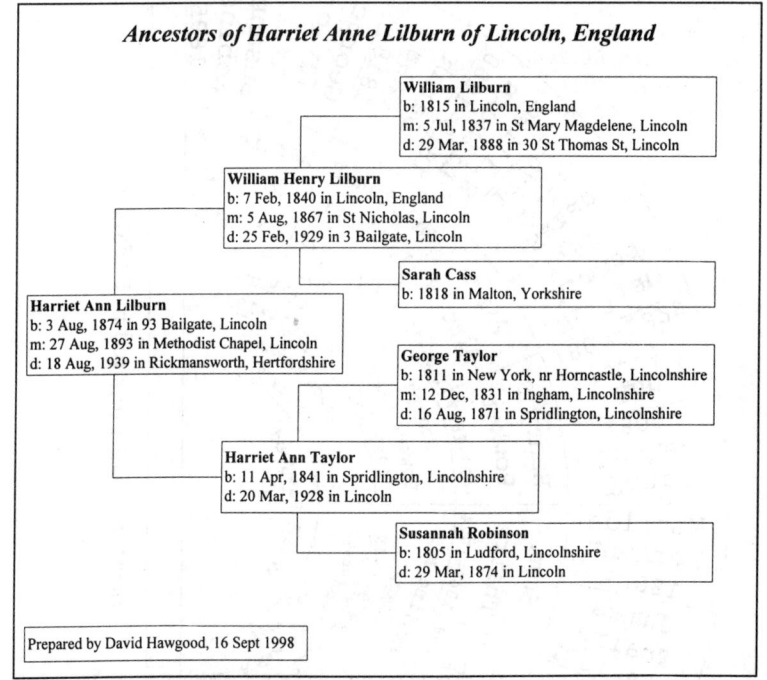

Prepared by David Hawgood, 16 Sept 1998

Figure 6. Ancestors chart, as prepared in Family Tree Maker

Printing a chart is usually a question and answer operation. First choose the type of chart, person from whom it starts, and number of generations. This defines the people to be included and how they are linked. Choose a layout, and which pieces of information are to be included. Usually there is a basic (default) layout at the start, so the user can just print a preset style without making any decisions except the person to start from. Choose whether to look at the chart on the display screen, print it immediately, or put it into a disk file. Using the last method, 'print to disk', the chart can be incorporated into a word processor document.

94   HUMAN PEDIGREE STUDIES

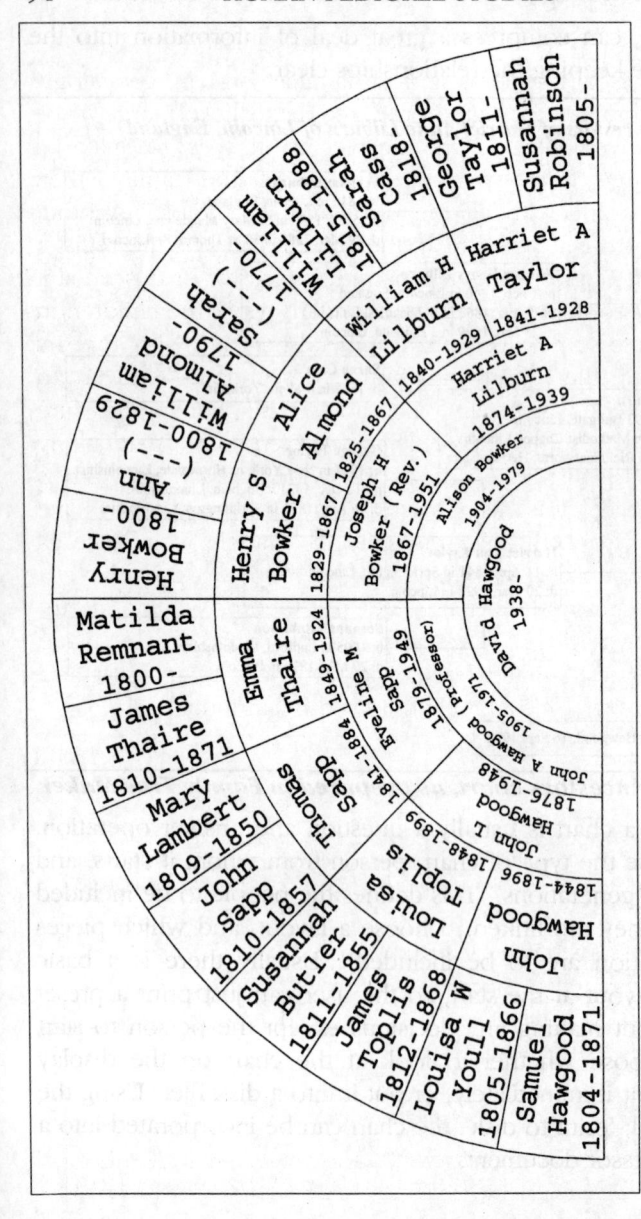

Figure 7. Fan style ancestors chart prepared in PAFMate

If there is a problem, it is a familiar one in genealogy. Putting a complete pedigree on a tree makes it too large to be printed conveniently. One answer to this is to make a summary tree to show the structure of the family, then individual charts with all the detail and footnotes. Another system, which is a standard in the United States, is the register format chart (Fig 8). This is a narrative ancestor or descendant chart with people numbered in a standard way - the children in a family have successive roman numerals, a child who appears further down the report with further descendants is given an arabic numeral as well. Footnotes and generations are numbered with superscripts. This style of report is particularly convenient for display on the World Wide Web. The width is limited, so display is easy. Even better, each person can have a hyperlink from the place as child in a family to a place as parent in a subsequent family. These reports usually have a page of footnotes and a name index at the end - again all are linked by hyperlinks. This type of report can also be made into an indexed printed book, with a variety of ancestor and descendant charts for a family followed by an index.

Descendants of George Taylor

Generation No. 1

1. GEORGE[1] TAYLOR was born 1811 in New York, nr Horncastle, Lincolnshire, England[1], and died 16 Aug, 1871 in Spridlington, Lincolnshire[2]. He married SUSANNAH ROBINSON 12 Dec, 1831 in Ingham, Lincs[3], daughter of ROBERT ROBINSON and ANN.

Children of GEORGE TAYLOR and SUSANNAH ROBINSON are:
     i. ROBERT[2] TAYLOR[4], b. 1831, Ingham, Lincs[4]; m. MARY.
     ii. MARY ANN TAYLOR[5], b. 1838[6].
2.   iii. HARRIET ANN TAYLOR, b. 11 Apr, 1841, Spridlington, Lincolnshire, England; d. 20 Mar, 1928, Lincoln.
     iv. GEORGE TAYLOR, b. 1847, Spridlington, Lincs[7]; m. MARY SMITH.

Generation No. 2

2. HARRIET ANN[2] TAYLOR *(GEORGE[1])*[8] was born 11 Apr, 1841 in Spridlington, Lincolnshire, England[8], and died 20 Mar, 1928 in Lincoln[9]. She married WILLIAM

HENRY LILBURN 5 Aug, 1867 in St Nicholas, Lincoln[10], son of WILLIAM LILBURN and SARAH CASS.

Children of HARRIET TAYLOR and WILLIAM LILBURN are:
      i.   GEORGE WILLIAM[3] LILBURN, b. 1869, Lincoln[11]; d. 1871, Lincoln[12].
      ii. HENRY TAYLOR LILBURN, b. 1871, Lincoln[13]; d. Oct 1871, Lincoln[13].
3.   iii. SUSANNA LILBURN, b. 1872, Lincoln; d. 1944.
4.   iv. HARRIET ANN LILBURN, b. 3 Aug, 1874, 93 Bailgate, Lincoln, England; d. 18 Aug, 1939, Rickmansworth, Hertfordshire, England.
5.   v.   ALICE LILBURN, b. Abt. 1879; d. 1944.

Generation No. 3

3. SUSANNA[3] LILBURN *(HARRIET ANN[2] TAYLOR, GEORGE[1])* was born 1872 in Lincoln[14], and died 1944[14]. She married WILLIAM BENNET ROBINSON 1896[14].

Children of SUSANNA LILBURN and WILLIAM ROBINSON are:
      i.   MARJORIE[4] ROBINSON, b. 18 Jun, 1897; d. 10 Jun, 1966.
      ii. BARBARA LILBURN ROBINSON, b. 1909; d. 5 Feb, 1991, Wimbledon, London.
      iii. RICHARD LILBURN ROBINSON, b. 1912; d. 1949.

4. HARRIET ANN[3] LILBURN *(HARRIET ANN[2] TAYLOR, GEORGE[1])* was born 3 Aug, 1874 in 93 Bailgate, Lincoln, England[15], and died 18 Aug, 1939 in Rickmansworth, Hertfordshire, England[16]. She married JOSEPH HENRY BOWKER 27 Aug, 1893 in Methodist Chapel, Lincoln[17], son of HENRY BOWKER and ALICE ALMOND.

Children of HARRIET LILBURN and JOSEPH BOWKER are:
      i.   HENRY ALAN[4] BOWKER, b. 5 May, 1896; d. 26 Dec, 1946, K Edward VII Hospital, Windsor, Berks; m. KATHLEEN STANHOPE LISTER, 30 Aug, 1923.
6.   ii. ALISON BOWKER, b. 18 Jul, 1904, Macclesfield, Cheshire; d. 16 Jul, 1979, Manchester, Lancs.

5. ALICE[3] LILBURN *(HARRIET ANN[2] TAYLOR, GEORGE[1])* was born Abt. 1879, and died 1944. She married FREDERICK MYERS Abt. 1904.

Children of ALICE LILBURN and FREDERICK MYERS are:
      i.   JEFFREY[4] MYERS, b. 1906.
      ii. CHARLES MYERS, b. 1908; d. 1929.

Generation No. 4

6. ALISON[4] BOWKER *(HARRIET ANN[3] LILBURN, HARRIET ANN[2] TAYLOR, GEORGE[1])* was born 18 Jul, 1904 in Macclesfield, Cheshire, and died 16 Jul, 1979 in Manchester, Lancs. She married JOHN ARKAS HAWGOOD 21 Dec, 1927 in British vice-consulate, Vienna, Austria, son of JOHN HAWGOOD and EVELINE SAPP.

Children of ALISON BOWKER and JOHN HAWGOOD are:
      i.   JOHN[5] HAWGOOD, b. 20 Jun, 1931, Maidenhead; m. MARY RUTH

> CLIBBON, 13 Jul, 1957, Stoke by Nayland, Suffolk.
>
> ii. ALAN HAWGOOD, b. 16 Feb, 1933, Maidenhead, Berks; d. 7 Jul, 1975, Formby, Lancs; m. VALERIE ANN SHEPPEY, 8 Oct, 1960, Birmingham.
>
> iii. DAVID HAWGOOD, b. 20 Jan, 1938, Kemerton, Worcs; m. BARBARA JEAN EXCELL, 27 Apr, 1974, Christ Church, Esher, Surrey.
>
> iv. CHRISTOPHER HAWGOOD, b. 9 Jun, 1941, Kemerton, Worcs; m. ELSE BARTELS, 31 Aug, 1985, Glostrup, Denmark.
>
> v. EVELYN ANNE HAWGOOD, b. 16 Mar, 1943; m. PARK OGBORNE, 9 Jul, 1982, Registry Office, Stockport.
>
> *Endnotes*
> 1. 1851 census Spridlington.
> 2. Death cert. of George Taylor.
> 3. pr of Ingham, Lincolnshire, 1831.
> 4. 1851 census Spridlington.
> 5. pr of Ingham, Lincolnshire, 1838.
> 6. pr of Ingham, Lincolnshire.
> 7. 1851 census Spridlington.
> 8. *Alison Hawgood's birthday book.*
> 9. death cert. of Harriet Ann Lilburn.
> 10. marr. cert. of William Henry Lilburn.
> 11. pr of St Mary Magdalene, Lincoln, 1871 burial, age 2.
> 12. pr of St Mary Magdalene, Lincoln, 1871 burial.
> 13. pr of St Mary Magdalene, Lincoln, 1871 burial, age 5m.
> 14. Family information from Mary Starling.
> 15. birth cert. of Harriet Ann Lilburn.
> 16. death cert. of Harriet Ann Bowker.
> 17. marr. cert. of Joseph Henry Bowker.

Figure 8. Register format descendants chart

Genealogy research using computers

I have mentioned that genealogy packages or databases can be used to select, sort and list information, to help decide which people are indeed members of one family. Some of this comes from searches in record offices. But in more and more cases the information, or an index to it, comes in computer form. I will give examples of computer-prepared indexes, data on disk, and data from the Internet.

Most UK family history societies have indexed the 1851 census for their counties using computers. This was done as relatively small individual projects. For example in Lincolnshire each census from 1841 to 1891 is indexed, published in booklets and on microfiche a registration district at a time. I have the book indexing the 1891 census for Lincoln (ref 5), and have used this and others in the series to find my Lilburn ancestors and relatives around Lincoln.

Quite different in scale is the International Genealogical Index, the IGI. Compiled by the LDS Church, this is an index mainly of births, baptisms and marriages from sources in many countries. It contains about 280 million entries. For the United Kingdom, most entries are from parish registers. The index was published on microfiche, county by county or state by state. It is now available in libraries on CD-ROM, one continuous alphabetic run for each country or region. It is part of a computer system called FamilySearch (Ref 6). The user can select entries and copy them to a floppy disk to take away. For example I have about 5000 Excell entries from the UK for my wife's maiden name Excell and similar names. They can either be copied in a word processor format, or in the GEDCOM format mentioned above. Using GEDCOM, I have copied all these Excell entries into a genealogy package, Pedigree. Fig 9 shows the result after I selected entries for St Botolph Bishopsgate among these 5000, sorted them into date order, and printed them in a word processor.

Christenings in St Botolph Bishopsgate, London from the 1988 IGI on CD-ROM			
Child	Father	Mother	Date
Paule	Austin Exall		30 Oct 1625
Mary	Auten Exall	Mary	23 Dec 1627
Alice	Austin Axall	Mary	26 Sep 1630
Ann	Austin Oxall	Mary	3 Nov 1633
Frances	Austin Oxall	Mary	17 Jan 1636

Figure 9. Entries from the International Genealogical Index

There are special purpose genealogy packages that help analyse information from the IGI, and display it on maps to make the distribution apparent. Fig 10 shows IGI information for the SAPP surname. This was processed by the package BIRDIE, selecting information for each century, and producing a

USE OF COMPUTERS FOR GENEALOGICAL RESEARCH 99

map with the counties of England coloured according to the number of entries. It appears to show that the surname was predominantly in Norfolk in the 16th century, then became strong in London and Sussex also, ending up with none in Norfolk and most in Sussex and Wiltshire. But some of this is an artefact. It happens that an enthusiastic researcher searched many of the registers of Norfolk and submitted the information to the IGI - but no one did the same for Sussex. The IGI is a mixture of submitted entries and ones systematically extracted from parish registers and similar - but in this case the submissions win, and give a false picture. There are only 200 entries altogether, there are 17 of the Norfolk 16th century ones.

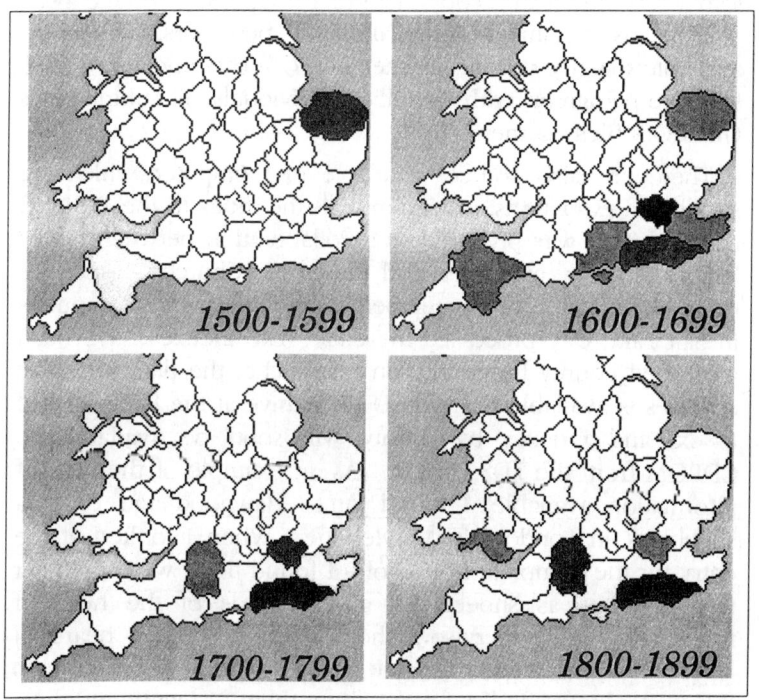

Figure 10. IGI information for surname SAPP, mapped by the package BIRDIE

But these are individual entries, not pedigrees. Within FamilySearch the LDS Church have another database, Ancestral File, which is a collection of submitted pedigrees, filling seven CDROM disks, with information on millions of individuals. When I started researching my Hawgood family the earliest ancestor I knew was a pawnbroker Samuel Hawgood from the Old Kent Road in London, born 1804. By systematic research, and correspondence with other Hawgoods, I found that his father came from Northamptonshire and have a definite line back another 100 years. I have submitted this information to the Ancestral File, so that the results of my research are preserved for the future and available to anyone worldwide.

There is a similar compilation of pedigrees, GENSERV, available directly on the Internet (ref 7). This contains 11,000 separate pedigrees, with 14 million individuals, all submitted as GEDCOM files via the Internet.

There are also sources available on computer which are systematic extractions. An example is the index to the 1881 UK census. This was prepared as a collaboration between family history society members and the LDS Church - societies transcribed, LDS Church members entered data, LDS computers in Salt Lake City processed the data. The indexes have been published county by county on microfiche, the data for some counties is available from the Data Archive at the University of Essex, and indexes by county will soon be available on CDROM from the LDS Church. As an example of the type of information available, I found James Hawgood age 16 as a shoeblack in London, in Marylebone. When I looked at the entry for the complete household, I found there were 32 other teenage boys as shoeblacks, all as wards of the head of household, and supervisor, Thomas Scarfe. I had heard of London's "Boot Black Boys" but had not expected to find them this way - I must find out more about them. I mentioned them in an email message on a society mailing list, and was told about William Quarrier, philanthropist, who set up a similar

USE OF COMPUTERS FOR GENEALOGICAL RESEARCH

organisation in Glasgow after 1864. So this chance find of James Hawgood as a shoeblack has set off a correspondence - this is quite normal in Internet discussions. Incidentally, I have extracted all Hawgood births up to 1881 from the General Register Office Indexes, and all Hawgood deaths. I aim to use my computer to correlate all births, marriages, deaths and 1881 census entries for these Hawgoods. For each death, I calculate the birth year from the age at death. I sort the information in various ways, for example by forename and birth date. Although ages are often a few years out, this at least gives a list of candidates who may be the same person. I think this shoeblack James Hawgood is the same as one who died age 34 in St Pancras, but I can find no registered birth which agrees with his 1881 census entry.

With an uncommon surname like Hawgood, it is fairly easy to correlate the different pieces of information, and it is feasible to follow up the entries I find in indexes and look at the original documents. This is much harder with common surnames - but indexes on computer are beginning to help here, particularly when the index entries give places and occupations as well as names. For example, I want to find more about my ancestor George Taylor. The family story is that he was a road contractor, but census entries show him as a turnpike road labourer. He was born in 1811 in New York - not the one in the USA, the tiny hamlet in the fens of Lincolnshire. Later he lived in the village of Spridlington outside Lincoln. I knew he was alive at the time of the 1871 census, and could not find him in the 1881 census. I could not face searching for the record of his death. But the Lincolnshire Archives have published various personal name indexes on CDROM - and I was pleased to find a reference to the 1871 will of George Taylor of Spridlington. The will showed him as a yeoman, from the probate I found his death certificate which showed him as a road contractor. But when his wife Susannah died her death certificate showed her as "road labourer's widow". This example illustrates the way indexes help find people with common surnames, but it also illustrates the discrepancies between different records.

Another way the computer helps research is by documenting the research process - searches to be done, searches which have been tried but produced nothing. Fig 11 is a "To Do" list from Family Origins - I have selected records to examine at Lincolnshire Archives. All of these result from index entries, one from the 1881 census index, two from the Lincolnshire section of the National Burial Index (in progress), one from Lincolnshire Marriage indexes. The Robinsons in this list are another example of the way computerised indexes are helping my search for ancestors with common surnames.

8. Lincolnshire Archives
Look for baptism 1805/6 Holton-cum-Beckering David Robinson
9. Lincolnshire Archives
detail of 1810 marr. St Swithin, Lincoln   John Lilburn
10. Lincolnshire Archives
detail of 30 Jan 1852 burial St Martin, Lincoln   John Lilburn
11. Lincolnshire Archives
detail of burial Spridlington 1836 Ann (Robinson)

Figure 11. Research tasks "To Do", report from Family Origins

Internet for Genealogy

In considering how computers are used for family history research, I come now to the types of information obtainable by computer through the Internet. Can I "do my family tree" by searching the World Wide Web? Generally, the answer is no - there are very few systematic sources of information available on the Internet. An exception is Scotland - Scots Origins (Ref 8) includes the GRO indexes for 1855-1897, Church of Scotland baptism and marriage indexes and Scottish 1891 census index.

USE OF COMPUTERS FOR GENEALOGICAL RESEARCH 103

There is a charge, £6 to download up to 30 screens worth of results, each screen carrying up to 15 references. An American source available free is the US Social Security Death Index. But so far these are the exceptions. I have mentioned GENSERV, thousands of pedigrees available for searching on Internet. But probably the biggest use of the Internet is individuals announcing their families of interest, and asking for information from other genealogists. It may sound random, but the volume of messages is enormous, and they can be searched systematically.

Fig 12 shows the result of a search on a list of surname interests, the Roots Surname list. Each entry has a surname; the dates of interest; places with standard 3-letter abbreviations for countries, states and counties; with the places, migrations shown by ">" symbols and occasionally by the name of the ship; and a contact code for the submitter - the list is followed by one giving email or postal addresses to go with these codes. Expanding the last entry in the table, it shows that a researcher with code "jstephens" has information about a Robinson family which was in Scotland in 1784, moved to Northumberland in England, then moved to New South Wales in Australia and was there in 1862. The September 1998 list has about half a million entries, submitted by 70,000 researchers (ref 9). There are other similar lists, organised for particular counties or regions.

ROOTS SURNAME LIST			
Robinson	1607	1900	ENG>"Goodspeed">USA vmatneyr
Robinson	1680	1700	YKS,ENG tiss
Robinson	1700	now	Wonton,HAM,ENG jppdsp
Robinson	1700	1830	Alnmouth,NBL,ENG julies
Robinson	1784	1862	SCT>NBL,ENG>NSW,AUS jstevens

Figure 12. Some Robinson entries from the Roots Surname List

There are Internet mailing lists for discussion of almost every area of interest in genealogy. The not-for-profit organisation Rootsweb hosts over 4,000 independently-operated mailing lists, and in the month of August 1998 sent 103 million pieces of e-mail to its subscribers. Rootsweb also hosts over 3000 websites for family history societies or individual genealogists - there are thousands more on other servers.

Rather surprisingly, this apparently random system does produce results in finding cousins. A Hawgood in Hong Kong found my name in a directory of email addresses, it turns out we are 7th cousins. I put a set of web pages showing my Hawgood ancestors on the site of the publisher of Family Origins, because I was demonstrating the package. Next day, someone from Sussex emailed me - and we turned out to have common ancestors in Northamptonshire.

One feature of these email discussions is that the wired-up genealogist is describing interesting ancestors as well as distinguished ones. A mailing list for Australian genealogy ran a competition - who's got the most convict ancestors? Lesley Albertson described his great-grandfather John BEASLEY. (Ref 10)

"He broke into Mr Lambert's store in Milton, Berkshire. He was tried at Reading in January 1841, sentenced to 10 years, and shipped to Tasmania. He is forever being confused with other people, notably a convict from Greenwich, also called John BEASLEY. The wrangle between me and my Beasley cousins involved studying the genetics of eye colour ('mine' had grey eyes, 'theirs' had brown) and contemplating an ultrasound of his grave to determine the length of his skeleton".

This shows the mass of detail that family historians acquire about their ancestors. Another message, sent to me because of my wife's Excell ancestry, also illustrates the attitudes. It is from Stuart Tamblin, who has published indexes of criminal records (Ref 11). He had found:

John EXILE, Lent Assizes, Buckingham 1812; horse stealing: Death

Joseph EXILE, Lent Assizes, Buckingham, 1812; horse stealing: Death - Executed

USE OF COMPUTERS FOR GENEALOGICAL RESEARCH

He finished his message to me by saying: "Hope they're yours!". The peak of genealogy now is to find an ancestor who stole a horse, and was executed for doing it.

Notes and References:

The author will put links to web sites mentioned onto his own web pages, see http://ourworld.compuserve.com/homepages/David_Hawgood/ or http://www.hawgood.co.uk

Ref 1. Genealogy software packages are reviewed and their suppliers are listed regularly in the journal Computers in Genealogy published by the Society of Genealogists, 14 Charterhouse Buildings, London EC1M 7BA, England; the most recent listings are "Windows Genealogy Software" by Eric D Probert, vol 6 no 5 (March 1998) p225-232, and "Genealogy Software" by Eric D Probert, Vol 6 no 6 (June 1998) p266-273. A software directory is also published annually in the journal Genealogical Computing, PO Box 476, Salt Lake City, UT 84110, USA; see "1997 Directory of Genealogy Software", vol 17 no 2 (Oct/Nov/Dec 1997), p35-44. There are also listings in Family Tree Magazine, 61 Great Whyte, Ramsey, Huntingdon, Cambs PE17 1HL, England, for example "Commercial Genealogy Software" by Trevor Rix, vol 14 no 4 (July 1998) p32. These journals also have regular announcements of new or changed Internet addresses of relevant Web sites. Books listing or comparing genealogy packages include An Introduction to ... Using Computers for Genealogy by David Hawgood, 2nd edn, ISBN 1 86006 081 1, (Federation of Family History Societies, Ramsbottom, 1998); Computer Genealogy Update by David Hawgood, ISBN 0 948151 14 5, published by the author (London 1997); CAGe: Computer Aided Genealogy by Nigel Bayley, 2nd edn, ISBN 1 86150 007 6, (S & N Publishing, Salisbury 1998); Computers in Genealogy Beginners' Handbook edited by Neville Taylor, 2nd edn, ISBN 1 85951 017 5, (Society of Genealogists, London 1996).

Ref 2. "GEDCOM Data Transfer", by David Hawgood, published by the author, 2nd edn ISBN 0 948151 09 9, (London 1994).

Ref 3. The GEDCOM Standard Release 5.5, by the Family History Department of The Church of Jesus Christ of Latter Day Saints, 50 East North Temple Street, Salt Lake City, UT 84150, USA, published Dec 1995. For an electronic version of the GEDCOM Standard 5.5 see ftp site ftp://gedcom.org/pub/genealogy/gedcom

Ref 4. PAFMate published by Progeny Software in Canada became part of Corel Family Suite, which then became Family Heritage, published by IMSI.

Ref 5, 1891 Census of Lincolnshire, Index of Surnames. Vol 8, Lincoln Registration District, RG12/2587-2596, by Members of the Lincoln Family History Group, ed E & E Emptage, ISBN 1 872584 80 2 (Lincolnshire Family History Society, 1992).

Ref 6. FamilySearch, produced by the Church of Jesus Christ of Latter Day Saints, is available on computers in their Family History Centres worldwide, and many other libraries.

Ref 7, GENSERV, Internet Web address http://www.genserv.com. This and other collections are described in "Searchable Lineage Linked Databases" by Alan Mann, Genealogical Computing Vol 17 no 3 (Jan/Feb/Mar 1998) p11-13.

Ref 8, Scots Origins, Internet Web address http://www.origins.net/GRO. The information is provided by the Registrar General For Scotland, New Register House, West Register St, Edinburgh EH1 3YT, Scotland.

Ref 9, Roots Surname List, Internet web address http://www.rootsweb.com/rootsweb/searches/rslsearch.html. This and other surname data resources are described in "Finding Worth on the Web" by George Archer, Genealogical Computing Vol 16 no 3 (Jan/Feb/Mar 1997) p20-24.

Ref 10, message by Lesley Albertson (email address <albertsn@alphalink.com.au>) on the Conference of Australia History mailing list, email addrress AUSTRALIA-D-Request@Rootsweb.com, digest V98 #18 of 9 Jan 1998. The story of the convict John Beasley is also given under the ship name (Tortoise), port of arrival (Van Diemen's Land) and date of arrival (19 Feb 1842) on the Internet Web site http://carmen.murdoch.edu.au/community/dps/convicts/stories.html#1. A copy of the story and references will be deposited in the Society of Genealogists Document Collection, under surname Beasley.

Ref 11, HO 27 Criminal Register Indexes are published on microfiche and disk by Stuart Tamblin, 14 Copper Leaf Close, Moulton, Northampton NN3 7HS, England; Web address http://ourworld.compuserve.com/homepages/Stuart_Tamblin. The Exile entries quoted are from Volume 8, Bucks and Herts 1805-1816.

Social, Ethical and Technical Implications of Pedigree Construction: What The Maps Tell Us About the Mapmakers[1]

Robert G Resta

The pedigree is the consummate scientific tool - universal, succinct and objective. Its simple "language" of circles, squares and lines is easily recognisable to geneticists anywhere. By conveying complex information that would otherwise require extensive explanation, a pedigree is literally the picture worth a thousand words.

Pedigrees usually provide clues and information about genetic relationships, gene mapping and recurrence risks. Sometimes, if read properly, pedigrees can also yield insight into the all-too-human geneticists who construct them.

In this paper, I will analyse a well-known pedigree to show how it can provide information about the scientific biases of the person(s) who created it.

The Wedgwood-Darwin-Galton Pedigree

The Eugenics Record Office in the United States and the Eugenics Education Society in England were the leading eugenics advocates in the English-speaking world in the early decades of the twentieth century. Among their many activities, these institutions constructed thousands of pedigrees of "defective" families in an attempt to show how heredity contributed to a variety of social problems such as crime, poverty, alcoholism, and feeble-mindedness.

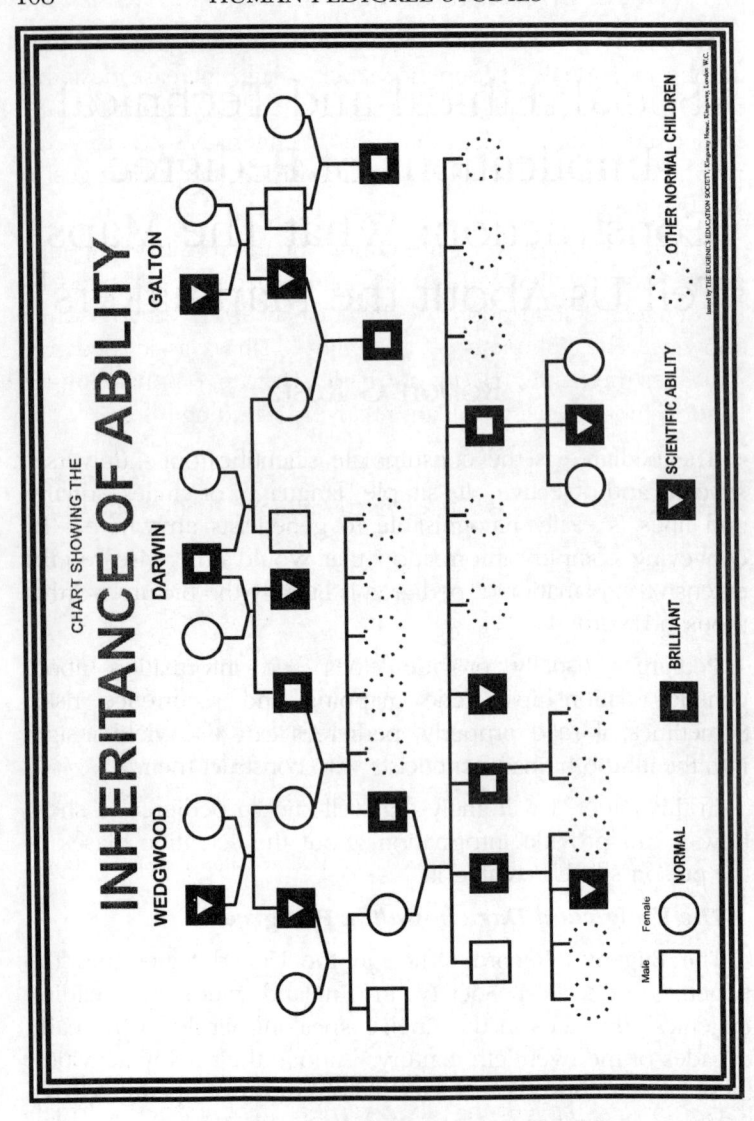

The Wedgwood-Darwin-Galton Pedigree

SOCIAL, ETHICAL AND TECHNICAL IMPLICATIONS

Conversely, eugenicists argued that encouraging breeding among individuals of "superior stock" could increase desirable traits such as intelligence, scientific ability, creativity, and leadership skills. To that end, eugenicists constructed pedigrees of select families to demonstrate the advantages of positive eugenics.

Perhaps the best known example of this positive eugenics genre is the Wedgwood-Darwin-Galton pedigree prepared by the Eugenics Education Society circa 1911, titled "Chart Showing The Inheritance of Ability." This classic pedigree illustrates the family history of Charles Darwin, emphasising his relationship to several brilliant relatives, including:

- Francis Galton, the official founding father of eugenics

- Josiah Wedgwood, founder of the famous Wedgwood china manufacturing company

- Erasmus Darwin, one of England's finest physicians in the late 18th century

I do not wish to dispute the impressive achievements of this remarkable family. However, the chart does not tell the entire story of the Darwin family. By closely examining this pedigree, I will show how these seemingly objective symbols tell a story that either was consciously manipulated or was filtered by a powerful eugenic lens. I have arbitrarily divided this analysis into two components – "The Brilliant Male Bias" and "Omitted Traits and Relatives."

The Brilliant Male Bias

The first unusual feature of this pedigree is the apparent inheritance pattern of the two traits in question. Both "Brilliance" and "Scientific Ability"[2] are manifested by, and inherited through, males only. Females do not possess these traits, nor are they even "silent carriers." For example, Josiah Wedgwood's son gets his intelligence from his father. Francis

Galton inherited his brilliance from his paternal ancestors, even though he is *maternally* related to the great Erasmus Darwin.

In a more subtle demonstration of a male-centred view of the inheritance of desirable traits, the sex of some "normal" males is masked. As the legend at the bottom indicates, squares denote males and circles indicate females. However, another category of individuals, "other normal children," is indicated with dashed circles. Unless the pedigree is carefully studied, one would assume that these "normal children" are female since circles typically denote females. For example, Darwin had four sisters and one brother (Erasmus Darwin) yet it would appear by looking at the pedigree that Darwin had five sisters. Furthermore, Darwin had 6 sons and four daughters. His last son, Charles Waring Darwin, is denoted by a hatched circle. Neither Darwin's son nor his brother were particularly brilliant nor did they have special scientific ability (Indeed, see below for more on Darwin's youngest son). In light of the pedigree's male bias, it is not surprising that these pedestrian males are "disguised" as females.

Not only are some males masked as females, many unremarkable males are omitted from the pedigree altogether. Undoubtedly, the entire pedigree could not be conveniently viewed in a concise format. But there is some pattern to the way that males are left out of the pedigree. As the pedigree stands, an impressive 16 of 20 (80%) men in the pedigree are "brilliant" or have "scientific ability". If the less remarkable males are included, then less than 50% of the men possess these traits, thus diluting the apparent heritability of these traits.

Omitted Traits and Omitted Relatives

When depicting families with undesirable traits such as feeble-mindedness, illegitimacy, and socially unacceptable behaviours, eugenicists often went to great pains to describe the shortcomings of the entire kindred. However, these same

SOCIAL, ETHICAL AND TECHNICAL IMPLICATIONS 111

negative traits were ignored in the Darwin family. For example:

- Erasmus Darwin stuttered markedly[3]. He fathered two illegitimate children in between his marriages, yet neither child is included in the pedigree[4]. Erasmus's son, also named Erasmus, committed suicide when he was 40 years old[5]. Erasmus Sr.'s first wife, Mary Howard, had a neuropsychiatric disturbance and used alcohol excessively[6]. Erasmus's second wife, Elizabeth Colyear, was herself the illegitimate daughter of Charles Colyear, 2nd Earl of Portmore[7].

- Charles Darwin suffered from a mysterious and debilitating lifelong illness that often left him incapacitated for long periods of time[8]. His father, Robert, suffered from gout[9]. Charles' youngest son, Charles Waring Darwin, died at 18 months of age, and was said to be mentally retarded[10]. Emma Darwin was 48 years old when she delivered this child, and so it is likely that the child had a chromosome defect, perhaps Down syndrome. In the pedigree, this boy is categorised with "other normal children."

- Francis Galton had a nervous breakdown as an adult[11]. His marriage to Louisa Butler never produced any children. Adele Galton, one of Francis's sisters, was often confined to a couch with "spinal weakness"[12]. His brother, Darwin Galton, had epileptic seizures.[13]

- Thomas Wedgwood, Josiah Wedgwood's brother, had multiple allergies, a mild learning disability and poor health[14]. Mary Ann Wedgwood, Josiah's youngest daughter, was mentally retarded and died at age eight[15]. Josiah's son Tom was addicted to opium and other drugs.[16]

- Emma Darwin's mother, Elizabeth Allen Wedgwood, had 8 sisters, five of whom were deaf[17]. Frances Julia "Snow" Wedgwood, novelist and great-granddaughter of Josiah, was also deaf[18].

- Inbreeding through cousin marriages was common in this extended family. For example, of Josiah Wedgwood's eight children, six eventually married. Of those six, four married cousins[19]. Only one cousin marriage (Charles and Emma) is depicted in the pedigree. Yet note this opinion from Ethel Elderton's eugenic tract *On The Marriage of First Cousins* "And, after an examination of the evidence .... we feel justified in asserting that in the bulk of cases cousin marriage is undesirable, even in those instances where the individual can boast of an apparently normal and healthy ancestry and collateral kinship." [20]

Family Secrets?

The dysgenic traits of Darwin's relatives were not skeletons hidden deep in the family's closet. Indeed, Charles Darwin himself was very concerned about the implications of his heritage for his children, as noted in this 1852 letter he wrote to his father, Robert: "My dread is hereditary ill-health. Even death is better for them [his children]."[21] Ironically, Robert Darwin expressed similar thoughts to his father more than 50 years earlier. Erasmus responded with a reassuring letter to Robert wherein he expounded on the heritability of the family's less desirable traits[22].

Nor is it likely that the Eugenics Education Society was ignorant of the family's woes. At the time the pedigree was produced, the president of the society was Leonard Darwin, the eighth child of Charles Darwin. Even if Leonard Darwin was reticent about his family's darker side, the trials and tribulations of the Darwin family are documented in standard biographical sources, as demonstrated by the ease with which I obtained most of this information from my local library.

There were other versions of the Darwin pedigree. The Eugenics Record Office in America produced a more extensive pedigree chart that listed virtually all of Darwin's relatives over 10 generations. However, this version neglected to include the family's eugenically undesirable traits. An extremely thorough pedigree produced by Karl Pearson claimed to trace Darwin's ancestry all the way back to Charlemagne! If so much genealogical information was available, then surely the family problems that I discussed above must have been known or readily available to anyone studying Darwin's lineage.

The analysis presented here demonstrates that even the most seemingly objective scientific tools and data can be strongly influenced by the psychological, social, and cultural profile of scientists. The scientific endeavour is carried out by human beings who are the products of their culture and their times. By acknowledging and studying the personal biases inherent in the scientific process, we can better appreciate the limitations and values of scientific information.

References:

[1] An earlier version of this paper appeared as Resta R, 1995, "Whispered Hints" Amer J Med Genet 59:131-133.

[2] It's not exactly clear what the distinction is between brilliance and scientific ability, and why Charles Darwin, one of the greatest scientists of all time, is considered "Brilliant" rather than having "Scientific Ability!"

[3] King-Hele D (1963) *Erasmus Darwin*. New York: Charles Scribner and Sons, p 16.

[4] Forrest DW (1974) *Francis Galton: The Life and Work of a Victorian Genius*. New York: Taplinger Publishing Company, p 1; King-Hele, p 22.

[5] King-Hele, 1963, p 37.

[6] Wedgwood B, Wedgwood H (1980) *The Wedgwood Circle 1730-1897: Four Generations and Their Friends*. London: Studio Vista/Cassell, p 46; King-Hele, 1963, p 21.

[7] King-Hele, 1963, p 31.

[8] Barlow N (ed) (1958) *The Autobiography of Charles Darwin, 1809-1882.* New York: Harcourt, Brace and Company; Bowlby J (1990) *Charles Darwin: A New Life.* New York: W.W. Norton; Smith F (1992) *Charles Darwin's health problems.* J Hist Biol 25, pp 285-306.

[9] Smith, 1992, p 300.

[10] Litchfield H, ed. (1915) *Emma Darwin – A Century of Family Letters 1792-1896, Volumes 1 & 2.* New York: D. Appleton, p 162.

[11] Forrest, 1974, pp 85-89.

[12] Forrest, 1974, p 5.

[13] Forrest, 1974, p 108.

[14] Smith, 1992, pp 292-293.

[15] Bowlby, 1990, p 33; Wedgwood & Wedgwood, 1980, pp 79,89.

[16] Bowlby, 1990, p 37; Wedgwood & Wedgwood, 1980, p 113.

[17] Wedgwood & Wedgwood, 1980, p 143.

[18] Wedgwood & Wedgwood, 1980, p 236.

[19] Brent P (1981) *Charles Darwin: A Man of Enlarged Curiosity.* New York: Harper and Row.

[20] Elderton EM (1911) *On The Marriage of First Cousins.* Eugenics Laboratory Lecture Series, IV. London: Dulau and Company, p 38.

[21] Smith, 1992, p 306.

[22] Bowlby, 1990, p 39; Barlow, 1958, pp 223-225.

Index

A

AIDS, 79
American Breeders Association, 24, 25, 77
American Journal of Human Genetics, 73
American Journal of Medical Genetics, 73
Ancestral File, 90, 100
ancestral paths, 50
Annals of Eugenics, 67
Arabian, 19
Association of Genealogists and Record Agents, 13

B

Bateson, William, 67
biometry, 22, 45
BIRDIE, 98, 99
Boyd's Marriage Index, 9
breeding isolate, 22
British Record Society, 11
Burke's *Peerage*, 4
Bynum, W F, viii

C

Carnegie Institution, 29, 31
Carr-Saunders, 69
Children Act, 1975, 7
Cold Spring Harbor, 25, 27, 31
College of Arms, 5, 10
colour blindness, 63
consanguineous marriages, 28
county histories, 4
cousin marriages, 112

D

Darwin, Charles, viii, x, 64, 109, 111, 112
Darwin, Erasmus, 109, 110, 111
Darwin, Leonard, 112
Darwinism, vii
Davenport, Charles B, 25, 26, 28, 31, 33, 70, 72, 73, 74, 76
Diploma in "Genealogy and the history of the family", 14
Dugdale, Robert, 74
Dugdale, Sir William, 3

E

Edwards, Anthony, 47
Eugenical News, 29, 33, 37
eugenics, 62, 64, 65, 70
Eugenics Education Society, 19, 65, 67, 68, 69, 74, 107, 109, 112
Eugenics International, 37
eugenics movement, vii
Eugenics Record Office, 27, 28, 31, 33, 35, 65, 70, 107, 113
Eugenics Review, 69
Eugenics Society, 29, 32, 37, 41
exchange loop, 56, 57

F

family history, 13
Family Tree, 8
Family Tree Maker, 86, 87, 90, 92, 93
FamilySearch, 98, 100, 106
Federation of Family History Societies, 8, 15

Fisher, R A, 23, 32, 37, 45, 46, 47, 49, 67
foundation stock, 19
founder effect, 48, 49
Fragile X syndrome, 67

G

Galapagos Islands, 23
Galton Institute, v
Galton Laboratory, 65
Galton, Francis, vii, viii, ix, 18, 47, 48, 64, 66, 68, 76, 109, 110, 111
Garter King of Arms, 1, 7
GEDCOM, 89, 90, 98, 100, 105
Genealogical Society of Utah, 9, 10, 11
General Register Office, 101
genetic isolates, 49, 56, 57
genetic linkage, 46, 54, 58
genetic markers, 58
GENSERV, 100, 103, 106
Goddard, Henry, 69, 72
Gotto, Sybil, 67
Guild of One-Name Studies, 11

H

Harleian Society, 6
Harriman, Mary, 27
Hereditary Genius, viii, xi
Human Genome Project, 58, 59, 78
Huntington disease, 63, 79
Hutterites, 55, 57

I

identical by descent (IBD), 51, 52, 53, 54, 55, 59, 60
IGI, 98, 99
inbreeding loop, 55, 56

Institute of Heraldic and Genealogical Studies, 13
International Eugenics Committee, 41
International Federation of Eugenics Organisations, 41, 75
International Genealogical Index, 9, 98
Internet, 1, 89, 90, 97, 100, 101, 102, 103, 105, 106
Italian Eugenic and Genetic Society, 77

J

Jesus Christ of Latter-Day Saints, Church of, 9, 100
Journal of Genetics, 67
Journal of Heredity, 73
Journal of Medical Genetics, 67

L

Laughlin, Harry H, 28, 31, 33, 34, 35, 37, 38, 75
le Neve, Peter, 3
Lidbetter, E J, 20, 21, 22, 23, 26, 28, 31, 35, 41
linkage analysis, 50
London Zoo, ix
Lundborg, Hermann, 32

M

marriage ring, 57
Medical Research Council, 24
Mendel, Gregor, vii, viii, x, xi, 45, 46
Mendelian ratios, 68
Mendelism, 22, 24, 28, 31, 32, 33, 35, 38, 48, 49, 53
Mormon pedigrees, 58
Mormons, 9, 89
mutations, 51

INDEX

N
National Burial Index, 9, 102

O
one name study, 88
one-name search, 6
Open University, 14
Origin of Species, vii

P
pangenesis, viii, ix
Parish Register Society, 6
Pearson, Karl, 45, 66, 67, 68, 76, 113
Pedigree, 91
periodic table, vii
Plea Rolls, 2
probate records, 10
Public Records Act 1958, 14

R
register format chart, 95
regression theory, x
Rockefeller Foundation, 34
Roots Surname list, 103
Rootsweb, 103, 106
Royal Commission on the Poor Law, 19
Royal Horticultural Society, 24
Royal Society, 19
Rüdin, Ernst, 33, 34, 35, 36, 37, 76

S
segregation ratios, 46
Smith, John Maynard, x
Society of Genealogists, 1, 6, 7, 8, 9, 10, 11, 12, 13, 15
SpanSoft, 91

T
The American Breeders Magazine, 25
tombstone inscriptions, 10
TreeDraw, 91

U
University College, 65
University of Essex, 100
US Department of Agriculture, 24

W
Wedgwood, Josiah, 109, 111, 112
Weinberg, Wilhelm, 31, 32, 34, 35
Williams, Roger, 58
World Wide Web, 89, 95, 102

ALSO AVAILABLE IN THIS SERIES

MARIE STOPES, EUGENICS AND THE ENGLISH BIRTH CONTROL MOVEMENT

EDITED BY ROBERT A PEEL

Proceedings of the 1996 Conference of the Galton Institute

CONTENTS

Notes on the Contributors

Editor's Preface
Robert Peel

Introduction
John Peel

The Evolution of Marie Stopes
June Rose

Marie Stopes and her Correspondents: Personalising population decline in an era of demographic change
Lesley Hall

The Galton Lecture: "Marie Stopes, Eugenics and the Birth Control Movement"
Richard Soloway

Marie Stopes and the Mothers' Clinics
Deborah Cohen

"Marie Stopes: Secret Life" – A Comment
John Timson

Marie Stopes International Today
Patricia Hindmarsh

Index

ISBN 0950406627

Available post paid from the Institute's General Secretary Price £5.00

ALSO AVAILABLE IN THIS SERIES

ESSAYS IN THE HISTORY OF EUGENICS

EDITED BY ROBERT A PEEL

Proceedings of the 1997 Conference of the Galton Institute

CONTENTS

Notes on the Contributors

Editor's Preface
Robert Peel

The Theoretical Foundations of Eugenics
Greta Jones

Eugenics: The Early Years
Geoffrey Searle

Women, Feminism and Eugenics
Lesley Hall

From Mainline to Reform Eugenics - Leonard Darwin and C P Blacker
Richard Soloway

The Eugenics Society and the Development of Demography in Britain: The International Population Union, the British Population Society and the Population Investigation Committee
Chris Langford

Human Genetics
John Timson

Ninety Years of Psychometrics
Paul Kline

The Galton Lecture: "The Eugenics Society and the Development of Biometry"
Anthony Edwards

Eugenics in France and Scandinavia: Two Case Studies
Alain Drouard

Eugenics in North America
Daniel Kevles

Index

ISBN 0950406635

Available post paid from the Institute's General Secretary Price £5.00